Springer-Verlag 6900 Heidelberg 1 · Postfach 1780
Telefon (06221) 49101 · Telex 04-61723
1000 Berlin 33 · Heidelberger Platz 3
Telefon (0311) 822001 · Telex 01-83319

Springer-Verlag New York, NY 10010 · 175, Fifth Avenue
New York Inc. Telefon 673-2660

36 Fortschritte der chemischen Forschung
Topics in Current Chemistry

New Methods in Chemistry

 Springer-Verlag Berlin Heidelberg GmbH 1973

ISBN 978-3-540-06098-7   ISBN 978-3-540-38115-0 (eBook)
DOI 10.1007/978-3-540-38115-0

Contents

X-Ray Photoelectron Spectroscopy

Dr. Wolfgang Bremser

Application Laboratory, Varian GmbH, Darmstadt*

Contents

* Present address: Hauptlaboratorium B 9, BASF, Ludwigshafen (Rh.)

1. Historic Introduction

Photoelectron spectroscopy with x-ray excitation was virtually unknown before the appearance of Siegbahn's famous blue book [1], which became bibliophilic rarity due to the great interest of the chemical community in this new analytical technique. However, in the middle of the nineteen twenties H. Robinson [2] in England and M. de Broglie [3] in France studied the energie distribution of photoelectrons emitted by x-ray excitation. The photographically recorded spectra displayed characteristic long tails with sharp edges at the side of high kinetic energies. These edges were employed for determining the binding energies of electrons in different orbitals, giving only a rather crude value of low accuracy. The sharp and well defined photoelectron lines on the other hand resulting from the unscattered electrons could not be detected because of the limited resolution of the early spectrometers.

K. Siegbahn succeeded with his knowledge of beta-ray spectroscopy [4] to improve this resolution considerably and demonstrated that the narrow photoelectron line is not disturbed by the energy absorption processes leading to the long tails at lower energy. A more precise determination of binding energies permitted to detect the influence of changes in the chemical state of elements and the discovery of the "chemical shift" of the photoelectron line [5] pointed out the usefulness of these data for applications in chemistry. This probably prompted Siegbahn to name the new analytical technique "*E*lectron *S*pectroscopy for *C*hemical *A*nalysis" (ESCA) [6], a name still widely used because of its brevity, even though it is not very specific: the term "electron spectroscopy" is familiar also to scientists working in the uv-visible range, while most spectroscopic techniques are employed for chemical analysis.

Because of the rapidly growing interest, the development of commercially available spectrometers was necessary to provide the analytical chemist with an easy to use tool. J. Helmer [7,8] designed the first routine spectrometer with sufficient resolution and sensitivity and coined the name "*I*nduced *E*lectron *E*mission" (IEE). The preretardation of electrons [7] and the use of an electrostatic analyser instead of the famous Siegbahn double focussing iron-free magnetic spectrometer [9,10] enabled him to build a small, rugged and easy to handle system without the need for magnetic field compensation by Helmholtz coils [10]. Other commercial companies followed [11] and today the scientist has quite a large choice of instruments to fulfill the different requirements of his specific field of work.

The name most commonly used today is "*X*-Ray *P*hotoelectron *S*pectroscopy" (XPS) which clearly distinguishes ESCA from the closely related "*UV*-*P*hotoelectron *S*pectroscopy" (UPS). The latter technique uses UV-photons, normally a He(I)-lamp of 21.21 eV, for the liberation of photoelectrons. While only valence electrons can be observed, the resolution of UPS-spectra in the gas phase is much higher, thus yielding detailed information on the structure of molecular orbitals [12]. The development of this technique is closely connected with the names of D. Turner [13] and E. Heilbronner [14] and UPS-spectrometers can be found in many modern laboratories. However, because of the differences both in instrumentation [15] and in the information content of the spectra, we will restrict ourselves in the present article to the x-ray excitation and refer the reader interested in UPS to the appropriate review articles [12,16].

2. The Basic Experiment

As the name XPS implies, the analytical sample is irradiated with x-ray light and the energy distribution of the emitted photoelectrons is recorded. Fig. 1 shows a comparison between the four different processes which are caused by the interaction between x-ray photons and matter, and which can be observed by four different spectroscopic techniques [17,18].

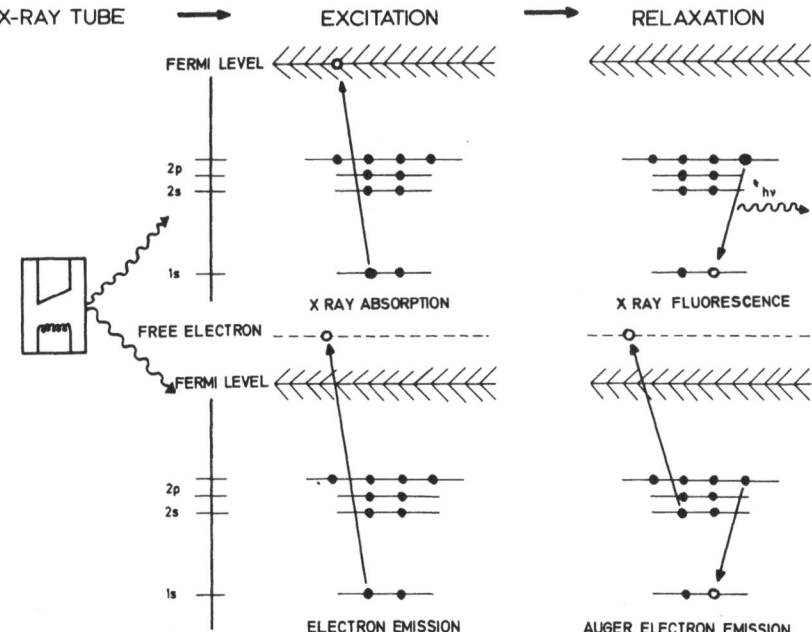

Fig. 1. Comparison of the four different physical processes which can be observed during the interaction of X-ray photons with matter [28]. The two phenomena scetched below, namely photoelectron emission and Auger electron emission, can be detected and measured in a photoelectron spectrometer by determining the kinetic energy of the ejected "free" electrons

3

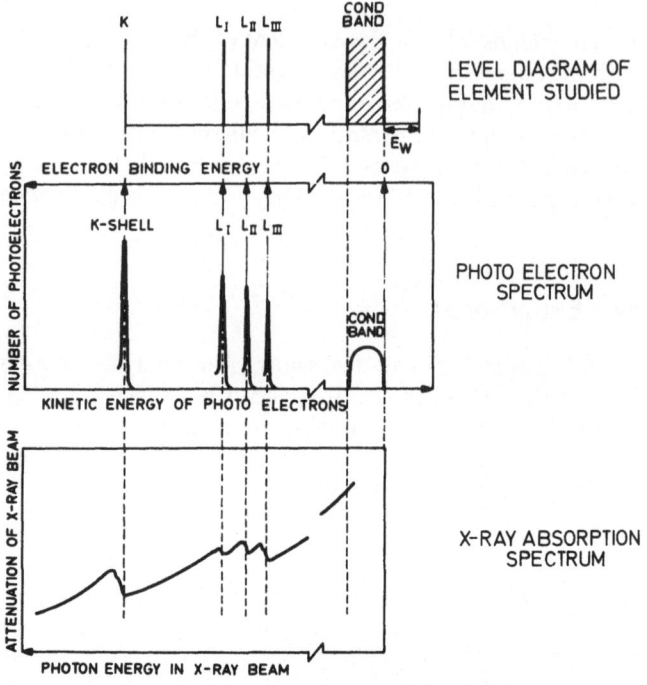

Fig. 2. Comparison of photo electron- and X-ray absorption spectrum [1]. Both types of spectra are a direct image of the orbital diagram of the element under study, however, the resolution in the X-ray absorption is much lower because of the numerous possible final states in the conduction or valence band

In the photoionisation process the x-rays liberate electrons in the various orbitals of the chemical compound. Depending on the experimental set-up and especially the photon energy $E_X = h\nu$, these electrons are either absorbed in the valence band near the Fermilevel, or they leave the atom as free electrons with a kinetic energy E_K,

$$E_K = E_X - E_I \tag{1}$$

which reflects the ionisation energy E_I of the orbital. While in x-ray absorption spectroscopy we measure the attenuation of the x-ray beam as a function of the frequency ν, photoelectron spectroscopy works with a constant photon energy which should be as monochromatic as possible. In principle both types of spectra contain the same information, however, the lines in the x-ray absorption spectrum are much broader due to the numerous possible electron states in the valence or conduction band (Fig. 2).

After the primary process we have a cation with a vacancy in one of its inner orbitals. In the following relaxation process this hole will be filled with an electron from an outer orbital. There are two possible ways in which the system can loose this surplus energy:

> either an x-ray photon is emitted (*x-ray fluorescence*) or
> a second photoelectron is ejected (*Auger electron emission*).

Which of these two competing relaxation mechanisms can be observed, is independent of the primary excitation and only a function of excited ion. For light elements the Auger process prevails [19], while the heavier atoms exclusively show fluorescence spectra.

Thus we have three possibilities to determine the electron binding energies by means of x-ray excitation:

1) from the edges of the broad x-ray absorption bands,
2) from the sharp lines in the x-ray fluorescence spectrum. However, this technique is restricted to elements of high atomic number and because of the demanding experimental conditions to rather stable inorganic compounds. Furthermore we only observe differences between two orbital energies which increases the line width, makes the whole spectrum more difficult to grasp and obscures chemical shift effects [20]
3) from the kinetic energy of the emitted photoelectrons, which either originate from the primary photoionisation process or from the Auger effect. Again the Auger process is confined to certain elements of the periodic table and three energy levels are involved, adding to the complexity of the spectrum and the observed line width [21]. However, chemical shifts can be observed [1] and increase the potential of this technique, especially for surface investigation [22].

Photoelectron spectra on the other hand are a direct image of the orbital structure of the element under investigation (cf. Fig. 2) because of the defined energy of the "free electron". This greatly facilitates the interpretation. All elements can be observed with approximately equal intensity [23,24]. Chemical shifts are clearly visible. At the same time, the intense Auger lines can also be detected by a normal XPS-spectrometer [25]. A distinction between these two types of photoelectrons is easily possible by varying the excitation energy E_x. Therefore, modern spectrometers contain a split dual anode for easy change-over from Mg Kα- to Al Kα-irradiation without mechanical alterations or loss in sensitivity.

3. The Photoelectron Spectrometer

The aim of the XPS-experiment is the determination of binding energies by measuring the velocity of the emitted photoelectrons. Fig. 3 shows the in-

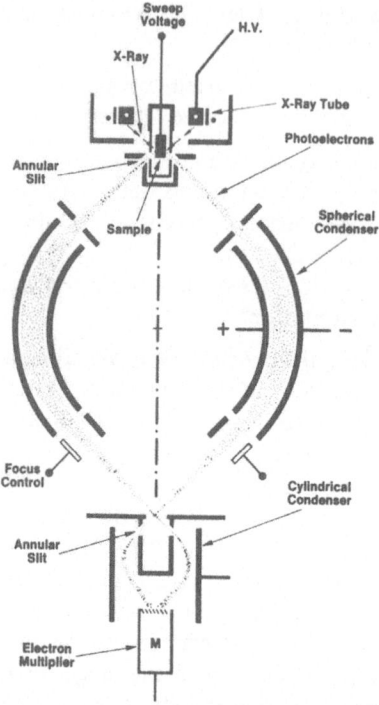

Fig. 3. Schematic presentation of a typical photoelectron spectrometer. The sample is introduced from the top. For detailed discussion see text

strumental realization of this principle. The whole arrangement is of rotational symmetry along the dotted vertical axis. This 360° angle of acceptance of the analyzer guarantees optimum sensitivity of the spectrometer for cylindrical samples and a high flexibility when different sources from different angles have to be incorporated. Such an arrangement with an electron gun for Auger spectroscopy, an Al- and Mg-x-ray source for XPS and an He(I)-lamp for UPS would allow an elegant combination of various techniques in one single spectrometer as well as "two-beam" experiments [30].

Electrons from the heated tungsten filament are accelerated to the annular anode. Depending on the anticathode material a characteristic fluorescence radiation is emitted, passes through a thin Aluminum window and induces photoelectrons on the surface of the analytical sample. These photoelectrons are deflected in the spherical electrostatic analyzer, double focussed to eliminate stray electrons and finally counted by the electron multiplier. The whole system works under a vacuum of 10^{-5} to 10^{-7} torr or even 10^{-10} torr, if surface properties have to be studied. This vacuum is generated by a Titanium sublimation- or a turbomolecular pump, the sample can be entered through a vacuum interlock without breaking the vacuum [27]. If ultra-high vacuum is desired or if volatile compounds are measured, provision for baking the system should be made in order to allow fast cleaning of a contaminated system.

The line broadening Δ_A introduced by an electrostatic analyzer is a function of the kinetic energy E_K of the photoelectrons

$$\Delta_A = E_K / A \tag{2}$$

where A is the resolution of the analyzer. In order to enhance the resolution without loosing too much sensitivity by changing the slit width, i.e. the value of A, the photoelectrons are retarded by applying a negative potential E_R to the sample and the sample cage [7]. This permits to analyze electrons at lower velocity

$$E_{K'} = E_K - E_R \tag{3}$$

and thereby higher resolution.

There are basically two different ways of sweeping a spectrum. Either the potential between the two spheres of the analyzer is increased continuously, thus bringing electrons of higher energy into the focus. Or the analyzer energy E_A, i.e. the kinetic energy of the electrons that are focussed at the specific ana-

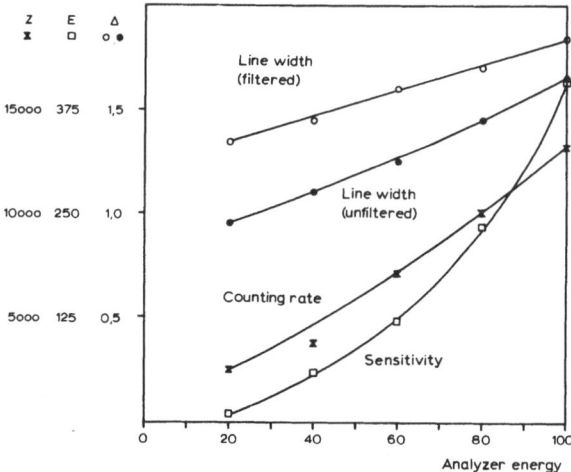

Fig. 4. Variation of line width (Δ), counting rate (Z) and sensitivity (E) with changing energy of the analyzed electrons [11]. It is apparent that the line broadening introduced by digital filtering is more severe for lower analyzer energies, i.e. higher resolution, because of the quadratic terms in Eq. 6. The increasing noise at lower kinetic energies is responsible for the non-linear decay of the signal to average noise ratio (E). An illustrative example is given in Fig. 12

lyzer potential, is kept constant and the retarding voltage is increased continuously. The latter technique, *i.e.* sweeping the retarding voltage, has the great advantage that both resolution and sensitivity are kept constant throughout the spectrum. Furthermore changing the analyzer energy between two experiments now provides an excellent possibility to either work under higher sensitivity and low resolution conditions or to trade off sensitivity in order to gain resolution by lowering the setting for the analyzer potential [28] (cf. Fig. 4). Because these variations of the operating parameters do not involve any mechanical adjustments like changing slits, they are practically continuous, easy to reproduce and do not need refocussing after each new setting.

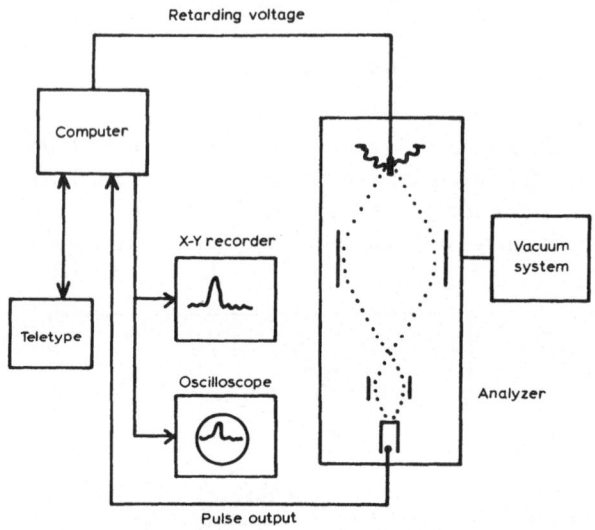

Fig. 5. Block diagram of a computerized photoelectron spectrometer [27]

In order to enhance the sensitivity [11] of the technique, time averaging over several scans is essential. Therefore, it was only a small step from the multichannel data storage device to a dedicated computer which can at the same time control the spectrometer and manipulate the raw data. Fig. 5 illustrates the flow of information [27] in a computerized system. The commands of the operator are given via teletype. The computer asks for all the experimental parameters needed, calculates the appropriate retarding voltage and starts scanning the desired spectral region. The pulses counted in the multiplier are stored in a specified part of the memory. Up to ten independent spectral regions can thus be analyzed and stored without an operator being present.

After the experiment the raw data can be inspected on the oscilloscope, manipulated and finally recorded on a flat bed recorder for further use. A teletype output of all the spectral parameters provides an excellent record for later reference [17].

There are several advantages of a spectrometer with an incorporated computer system. Besides the automated operation especially during overnight runs the computer offers a greater flexibility in the choice of spectral parameters. The accumulated data are not asymmetrically distorted by an analog filter, but can be digitally smoothed in a variety of ways [17] after the experiment without the risk of overfiltration. A digital output for further procession on a larger computer or for storage on magnetic tape is readily available. Spectra can be expanded or compressed for optimum presentation, more scans can be added for improving the signal to noise ratio and peak maxima can be located and counted for kinetic studies. Differentiation, integration and deconvolution subroutines allow further evaluation of the data. Examples of typical experiments are described in the literature [17,28].

4. Experimental Difficulties

In principle the determination of the electron binding energies E_B relative to the Fermi level (cf. Fig. 1) is done be measuring the kinetic energy of the emitted photoelectrons:

$$E_B = E_I - E_W = E_X - E_K - E_W. \tag{4}$$

The workfunction E_W is a spectrometer constant and represents mainly the work necessary to excite the electron from the Fermi-level to the free electron level. Bearing in mind the experimental set-up, where $E_A = E_{K'}$ is the constant analyzer energy, the complete equation reads

$$E_B = E_X - E_A - E_R - E_W. \tag{5}$$

Thus only the retarding potential E_R is varied in order to determine the binding energy E_B, the other three parameters are kept constant.

However, in practical work some difficulties are encounted and it seems appropriate to shortly discuss these problems in order to allow the reader a full appreciation of the results obtained and summarized in the second part of this article.

4.1 Sample Preparation and Handling

As mentioned above, the XPS-experiment is performed in a vaccum of 10^{-5} torr or better. If solid samples are to be examined, certain requirements concerning the vapour pressure of these compounds have to be met. If the volatility under operating conditions is too high, the sample should be cooled down. This can be done effectively by flowing cold nitrogen gas through the interior of the dewared probe holder, monitoring the gas temperature with a thermocouple and heating the nitrogen up to the desired temperature. The same arrangement also allows to reach working conditions well above room temperature. This variable temperature unit can be used to freeze down liquids if their vapour pressure is too high, however, it has to be pointed out clearly that the liquid state in itself is not a handicap, because a thin film of compound sticking on the surface of a sample holder will always be sufficient for the observation of a spectrum. This has been demonstrated with liquid polymers of low volatility.

On the other hand, with an efficient pumping system spectra of gaseous compounds can be observed, if the technique of differential pumping is employed. Gas spectra [29] normally show excellent resolution and no charging effects. Only water solutions as "liquid beams" [30] still cause severe instrumental problems and wait for a suitable solution.

There are numerous ways to prepare a solid sample for XPS-analysis [17]. Because of the 360° angle of acceptance, the optimum sample geometry is a cylinder of 11 mm diameter and 2 mm width. However, with a certain loss in sensitivity, flat samples as well as odd shaped samples can be analyzed as well, if they do not exceed the maximum size defined by the vacuum interlock. Sample powders are normally mounted on sticky tape or pressed into a metal grid [17]. Excellent results were also obtained by pressing pellets [31] or by subliming the sample directly unto the sample holder.

4.2 Surface Contamination

XPS is a surface technique. The photoelectrons leaving the atoms loose energy due to inelastic scattering, *i.e.* the plasmon excitations of the electron plasma in the specimen [32]. This effect results in the spectrum in a long tail at lower kinetic energies (higher binding energies), eventually summing up to the rising baseline observed in a survey spectrum (Fig. 9). However, because of the quantisation of this effect, some electrons leave the lattice without any energy losses and can be observed as the characteristic photoelectron line. The intensity ratio between line and tail, *i.e.* between unscattered and scattered electrons, depends on the kinetic energy of the electron and the characteristics of the lattice, but also on the distance of the excited atoms from the surface. The maximum escape depth of unscattered photoelectrons is about 50 to 100 Å for x-ray (Al Kα or Mg Kα) excitation [33,34] and much lower (two to three monolayers) when using UV light.

All this makes XPS a perfect technique for studying the surface properties of solid material. On the other hand, when trying to deduce bulk properties from the XPS-spectra of the outer layers, the spectroscopist has to ensure that bulk and surface are at least comparable, if not identical. Chemically speaking, the surface is often contaminated by oxidation, adsorption (mainly hydrocarbons, oxygen and carbondioxide), decomposition or crystal defeciencies, which have to be considered if reliable results are to be obtained. In many cases, simple exposure of the sample to the lab atmosphere is sufficient to influence the surface properties of the compound.

Several sample preparation techniques have been developed to overcome these problems. A reaction chamber flanged on top of the spectrometer allows all sort of sample treatments directly before the introduction into the analyzer chamber. Oxydation and reduction with different gases at various temperatures and pressures permit the simulation of reactions and the formation of a defined surface state. Sublimation in vacuo unto a cooled sample holder deposits a fresh surface. Sputtering, i.e. a gas discharge in argon atmosphere, cleans the surface from hydrocarbons and oxides [17]. Finally, the sample preparation in a glove box under nitrogen and the use of a transportation chamber for introduction of such an air-sensitive sample directly into the spectrometer avoiding any contact with the atmosphere adds to the versatility of the spectrometer.

4.3 Surface Charging

The workfunction E_W introduced in Eq. 5 actually contains three contributions:

the energy difference between the Fermi level and the free electron level. This effect should be small for gaseous compounds [29].

the potential difference between the compound and the entrance slit of the analyzer. Because of the unequal workfunctions of probe and slit, the resultant difference in macropotential creates an electric field which the photoelectron has to overcome with additional energy [35]. As long as this contribution does not change with time, it is a spectrometer constant and can be determined by calibration.

the phenomenon of surface charging of non-conductors [1,36,37], which can result in a chemical shift of more than 3 eV between materials of different conductivity [37,38].

Whereas the first two contributions are constant over longer observation times [39], the surface charge has to be determined for each measurement by observing the shift of a reference signal. Besides the ingeneous, but experimentally difficult approach of Perlman [36], several other techniques have been used in practise, e.g. the carbon surface contamination [1], the carbon of the sticky tape [24] or a standard added to form a homogeneous mixture [37]. Furthermore the surface charge also depends to a certain extent on the x-ray flux and the sample preparation [37]. Thus precise calibration is essential to obtain reliable values for the electron binding energies in insulators.

4.4 Resolution and Sensitivity

Like in all spectroscopic techniques resolution and sensitivity of the spectrometer determine the usefulness of an analytical technique and define the
limitations of its application.

Following Eq. 4 there are three different sources of line broadening adding to the natural line width Δ_B, which reflects the lifetime of the final state
(Heisenberg uncertainty principle) and in some cases an unresolved spin orbit
splitting. In first approximation [17]

$$\Delta^2 = \Delta_B^2 + \Delta_X^2 + \Delta_K^2 + \Delta_W^2 \tag{6}$$

only the influence of the largest parameters has to be considered. This is
normally not the limited resolution of the analyzer ($\Delta_K = \Delta_A$), because after
reducing the analyzer energy (Eq. 2) to 10 eV, the line width is only reflecting
the broadening of the x-ray source (Δ_X), the solid state effects (Δ_W), and the
orbital itself (Δ_B).

The non-monochromaticity of the x-ray photons results firstly from the
finite width of the $1s$- and $2p$-levels involved in the fluorescence process and
then from the overlap of the $K\alpha_1$-$K\alpha_2$-spin doublet. Not considering the
monochromatic UV-light [15], Magnesium and Aluminium with a line width
of 0.7 eV, resp. 0.9 eV proved to be most suitable as anticathode-material.
A further improvement is possible with the help of an x-ray monochromator [29,30], however, the inherent loss in sensitivity and the problems of probe
adjustment create a number of difficulties which still have to be overcome
in practise. Also the spectra obtained with the help of a monochromator
clearly indicate that the gain in resolution for the test standard graphite is
around 0.2 eV and the remaining 0.7 eV are obviously due to other parameters [30].

On the other hand, the monochromator has the great advantage that it
automatically eliminates the $K\alpha_{3,4}$-satellites. Fig. 6 demonstrates that this
can be achieved as well with the help of a Fourier transform filtering program, allowing a satellite free spectrum also for Magnesium irradiation where
monochromatization with the help of a spherically bent quartz crystal so far
has not been shown.

For solid materials the ultimate line width obtainable depends to a large
extent on the preparation of the sample. Smooth surfaces and clean metals
normally show only small broadening due to level bending or uneven charge
distribution. However, if the surface potential varies, especially if the total
charging is considerable [37], different atoms of the sample experience different workfuntions E_W. Besides there is the usual "solid state" broadening of
atomic levels into bands due to the matrix effects [30] and the crystal deficien-
cies on the surface.

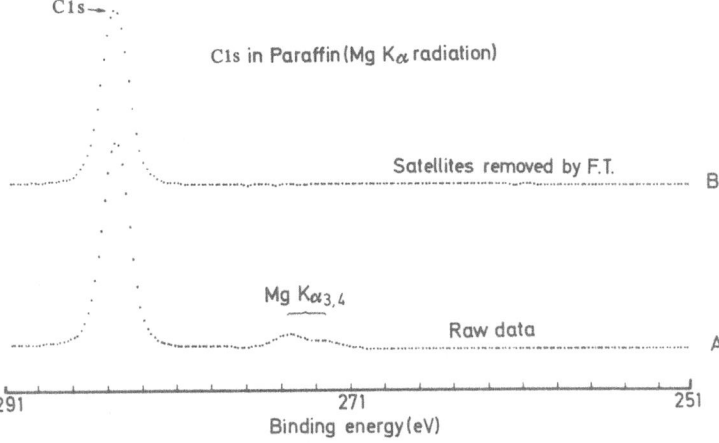

C1s

C1s in Paraffin (Mg Kα radiation)

Satellites removed by F.T.   B

Mg Kα₃,₄

Raw data   A

291   271   251

Binding energy (eV)

Fig. 6. Elimination of $K\alpha_{3,4}$-satellites in the C $1s$-spectrum of paraffin. The raw data in Fig. 6 A are transformed with the help of the Fourier transform filtering procedure into the satellite-free spectrum in Fig. 6 B. For details see example in Fig. 7

Summing up, there are several contributions to the line width, most of them well understood. However, like in all spectroscopic techniques, high resolution and high sensitivity are to a great deal mutually exclusive [11]. On the other hand, the improvement in sensitivity in the new spectrometers by means of a new anode design allows in many cases to trade off sensitivity for resolution. Another way to go is to employ the capabilities of the computer to artificially enhance the resolution. The classical way is the peak matching technique, where line position and intensity of two or more lines are varied until the sum of the overlapping lines gives an optimum fit with the experimental spectrum [17].

A more elegant and more flexible method is the Fourier transform deconvolution illustrated in Fig. 7. Spectrum 7a shows the chlorine $2p$-spin doublet in potassium chloride which is only partly resolved. This spectrum was Fourier analyzed, *i.e.* transformed from the frequency domain to the time domain (spectrum 7b). Deconvolution with the filter 7c resulted in the Fourier spectrum 7d, which was afterwards Fourier transformed into the frequency domaine spectrum 7e. A comparison of the initial spectrum 7a and the resolution enhanced spectrum 7e demonstrates the drastic gain in resolution accompanied by a slight loss in sensitivity. The great advantage of this technique is the possibility to construct a filter by deconvoluting two reference spectra or a spectrum and a curve obtained with the help of a mathematical function. This is the reason why these experimentally obtained filters can correct for very specific effects, like analyzer broadening, asymmetric x-ray broadening or x-ray satellites.

13

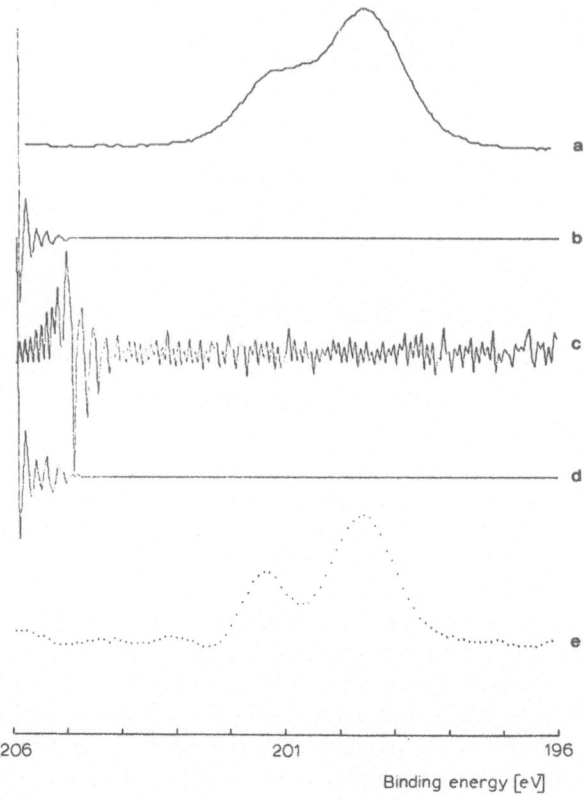

Fig. 7. Resolution enhancement of the chlorine 2p-spin doublet by Fourier transform deconvolution (see text for details)

5. Applications

Since the appearance of the first book solely devoted to XPS [1] many scientists started working in this field and have solved pending problems with the help of this new analytical technique. A number of review articles gave a more or less complete picture of the developments and helped to promote this unique yet still relatively unknown method. To allow a more general information I feel it worth mentioning some of the authors: K. Siegbahn [1,29,30,40,41], D. Shirley [42], R. S. Berry [43], D. M. Hercules [18,44], C. K. Jørgensen [24,45], J. M. Hollander [46], R. Nordberg [47], S. Pignataro [48], G. Mavel [49], W. Bremser [17,28], B. Lindberg [50] and C. Nordling [51].

5.1 Valence Bands in Solids

One of the first tasks of XPS was the precise determination of core electron binding energies for all elements of the periodic table. These data are now tabulated [1] and available for reference (Table 1). On the other hand, there is a great interest in the measurement of the range of low binding energies (0 – 20 eV) to get a clearer picture about structure and symmetry of the molecular orbitals.

The usefulness of UV-excitation for these experiments was already stressed on page 2. In many cases UPS and XPS give valuable complementary information [52], therefore an instrument for band structure analysis should permit different modes of excitation of both gases and solids. However, in the present article we have to restrict ourselves to studies on solids by means of x-ray photons.

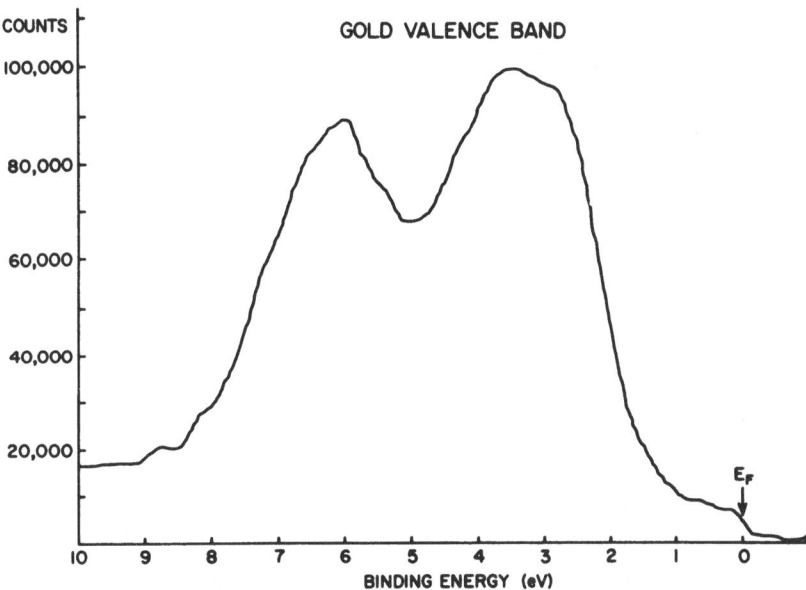

Fig. 8. Spectrum of the conduction band of metallic gold deposited on the sample holder by ion sputtering in the reaction chamber (Mg Kα irradiaton). Note the sharp step at the Fermilevel E_F

Conduction bands in metals represent at the moment the most interesting problems. Fig. 8 shows a spectrum obtained of the low energy range of gold. The small intensity s-band [30] extending to the Fermi level (E_F) is clearly visible. The spectrum itself is in good agreement with predictions by relativistic

Table 1. Reference list of core electron binding energies from 20 to 1100 eV

24	O $2s$	60	Co $3p$	141	As $3p_{3/2}$	331	Zr $3p_{3/2}$	546	Au $4p_{3/2}$
24	Sn $4d$	63	Ir $4f_{5/2}$	143	Pb $4f_{5/2}$	331	Pt $4d_{3/2}$	564	Ti $2s$
25	Ta $4f_{7/2}$	63	Na $2s$	149	Si $2s$	334	Au $4d_{3/2}$	571	Ag $3p_{3/2}$
25	Bi $5d_{5/2}$	67	Cd $4p$	158	Y $3d_{5/2}$	335	Pd $3d_{5/2}$	571	Hg $4p_{3/2}$
26	Ca $3p$	68	Ni $3p$	158	Bi $4f_{7/2}$	335	Th $4f_{7/2}$	572	Te $3d_{5/2}$
26	Y $4p$	69	Br $3d_{5/2}$	160	Y $3d_{3/2}$	340	Pd $3d_{3/2}$	575	Cr $2p_{3/2}$
27	Ta $4f_{5/2}$	70	Br $3d_{3/2}$	162	Se $3p_{3/2}$	344	Th $4f_{5/2}$	582	Te $3d_{3/2}$
27	Bi $5d_{3/2}$	70	Pt $4f_{7/2}$	162	Cs $4p_{3/2}$	347	Ca $2p_{3/2}$	609	Tl $4p_{3/2}$
29	Zr $4p$	73	Al $2p_{3/2}$	163	Bi $4f_{5/2}$	352	Au $4d_{3/2}$	617	Cd $3p_{3/2}$
29	Ge $3d$	74	Pt $4f_{5/2}$	164	S $2p_{3/2}$	360	Hg $4d_{5/2}$	620	I $3d_{5/2}$
31	F $2s$	74	Cu $3p$	180	Ba $4p_{3/2}$	363	Nb $3p_{3/2}$	628	V $2s$
31	Na $2p$	76	Tl $5p_{3/2}$	180	Zr $3d_{5/2}$	367	Ag $3d_{5/2}$	631	I $3d_{3/2}$
31	Hf $5p_{3/2}$	77	In $4p$	182	Br $3p_{3/2}$	373	Ag $3d_{3/2}$	641	Mn $2p_{3/2}$
32	Sc $3p$	77	Cs $4d_{5/2}$	182	Th $5p_{3/2}$	377	K $2s$	645	Pb $4p_{3/2}$
32	Sb $4d$	79	Cs $4d_{3/2}$	183	Zr $3d_{3/2}$	379	Hg $4d_{3/2}$	664	In $3p_{3/2}$
33	U $6p_{3/2}$	83	Au $4f_{7/2}$	188	B $1s$	380	Hf $4p_{3/2}$	677	Th $4d_{5/2}$
34	W $4f_{7/2}$	86	Pb $5p_{3/2}$	189	P $2s$	381	U $4f_{7/2}$	679	Bi $4p_{3/2}$
34	Ti $3p$	87	Zn $3p$	192	La $4p_{3/2}$	386	Tl $4d_{5/2}$	686	F $1s$
34	Nb $4p$	87	Au $4f_{5/2}$	195	U $5p_{3/2}$	392	U $4f_{5/2}$	695	Cr $2s$
35	Mo $4p$	88	Th $5d_{5/2}$	200	Cl $2p_{3/2}$	393	Mo $3p_{3/2}$	710	Fe $2p_{3/2}$
35	Re $5p_{3/2}$	89	Mg $2s$	205	Nb $3d_{5/2}$	399	N $1s$	714	Th $4d_{3/2}$
37	Ta $5p_{3/2}$	89	Sn $4p$	208	Ce $4p_{3/2}$	402	Sc $2p_{3/2}$	715	Sn $3p_{3/2}$
37	W $5p_{3/2}$	90	Ba $4d_{5/2}$	208	Nb $3d_{3/2}$	404	Cd $3d_{5/2}$	726	Cs $3d_{5/2}$
37	W $4f_{5/2}$	93	Ba $4d_{3/2}$	214	Hf $4d_{5/2}$	405	Ta $4p_{3/2}$	738	U $4d_{5/2}$
38	V $3p$	93	Bi $5p_{3/2}$	224	Hf $4d_{3/2}$	407	Tl $4d_{3/2}$	740	Cs $3d_{3/2}$
40	Te $4d$	95	Th $5d_{3/2}$	227	Mo $3d_{5/2}$	411	Cd $3d_{3/2}$	766	Sb $3p_{3/2}$
41	As $3d$	96	U $5d_{5/2}$	229	S $2s$	413	Pb $4d_{5/2}$	769	Mn $2s$
43	Cr $3p$	99	Sb $4p$	230	Mo $3d_{3/2}$	426	W $4p_{3/2}$	779	Co $2p_{3/2}$
43	Ru $4p$	99	Si $2p_{3/2}$	230	Ta $4d_{5/2}$	435	Pb $4d_{3/2}$	780	U $4d_{3/2}$
43	Th $6p_{3/2}$	99	La $4d$	239	Rb $3p_{3/2}$	438	Ca $2s$	781	Ba $3d_{5/2}$
45	Re $4f_{7/2}$	99	Hg $4f_{7/2}$	242	Ta $4d_{3/2}$	440	Bi $4d_{5/2}$	796	Ba $3d_{3/2}$
46	Os $5p_{3/2}$	103	Ga $3p_{3/2}$	246	W $4d_{5/2}$	443	In $3d_{5/2}$	819	Te $3p_{3/2}$
47	Re $4f_{5/2}$	103	Hg $4f_{5/2}$	259	W $4d_{3/2}$	445	Re $4p_{3/2}$	832	La $3d_{5/2}$
48	Rh $4p$	105	U $5d_{3/2}$	260	Re $4d_{5/2}$	451	In $3d_{3/2}$	846	Fe $2s$
49	Mn $3p$	110	Te $4p$	269	Sr $3p_{3/2}$	455	Ti $2p_{3/2}$	849	La $3d_{3/2}$
50	I $4d$	111	Be $1s$	270	Cl $2s$	461	Ru $3p_{3/2}$	855	Ni $2p_{3/2}$
50	Os $4f_{7/2}$	111	Ce $4d$	273	Os $4d_{5/2}$	464	Bi $4d_{3/2}$	875	I $3p_{3/2}$
51	Pd $4p$	111	Rb $3d_{5/2}$	274	Re $4d_{3/2}$	469	Os $4p_{3/2}$	884	Ce $3d_{5/2}$
51	Ir $5p_{3/2}$	112	Rb $3d_{3/2}$	279	Ru $3d_{5/2}$	485	Sn $3d_{5/2}$	902	Ce $3d_{3/2}$
51	Pt $5p_{3/2}$	114	Ce $4d$	284	C $1s$	494	Sn $3d_{3/2}$	926	Co $2s$
52	Os $4f_{5/2}$	118	Al $2s$	284	Ru $3d_{3/2}$	495	Ir $4p_{3/2}$	931	Cu $2p_{3/2}$
52	Mg $2p$	118	Tl $4f_{7/2}$	290	Os $4d_{3/2}$	496	Rh $3p_{3/2}$	968	Th $4p_{3/2}$
54	Au $5p_{3/2}$	122	Ge $3p_{3/2}$	294	K $2p_{3/2}$	500	Sc $2s$	998	Cs $3p_{3/2}$
55	Li $1s$	122	Tl $4f_{5/2}$	295	Ir $4d_{5/2}$	513	V $2p_{3/2}$	1008	Ni $2s$
56	Fe $3p$	123	J $4p$	301	Y $3p_{3/2}$	519	Pt $4p_{3/2}$	1021	Zn $2p_{3/2}$
56	Ag $4p_{3/2}$	133	Sr $3d_{5/2}$	307	Rh $3d_{5/2}$	528	Sb $3d_{5/2}$	1045	U $4p_{3/2}$
57	Se $3d$	135	P $2p_{3/2}$	312	Rh $3d_{3/2}$	531	Pd $3p_{3/2}$	1063	Ba $3p_{3/2}$
58	Hg $5p_{3/2}$	135	Sr $3d_{3/2}$	312	Ir $4d_{3/2}$	532	O $1s$	1072	Na $1s$
60	Ir $4f_{7/2}$	138	Pb $4f_{7/2}$	314	Pt $4d_{5/2}$	537	Sb $3d_{3/2}$	1096	Cu $2s$

band structure calculations [61]. Other studies on conduction bands and density of states in transition metals are reported by Y. Baer *et al.* [53,54] and in rare earthsmetals by S. Hagström *et al.* [55,54]. Detailed work on organic heterocycles [57], on graphite [58] and its isoelectronic compounds [59] as well as on some lithiumsalts [60] has been published recently. Even though there are several contributions to the photoelectron spectrum, that make it a non-trivial task to extract a true picture of the density of states, this specific application is probably the most promising for the theoretician.

5.2 Qualitative Analysis

Knowing the binding energies of core level electrons of the elements commonly present in a chemical compound, these data can be arranged with increasing magnitude (cf. Table 1) and used for the identification of photoelectron peaks in an unknown compound. A survey spectrum of potassium alaun KAl $(SO_4)_2$ in Fig. 9 clearly shows the different elements present. If two energy values in the table lie close together, an exact distinction can be made with the help of the spin-orbit splitting with its characteristic value for each

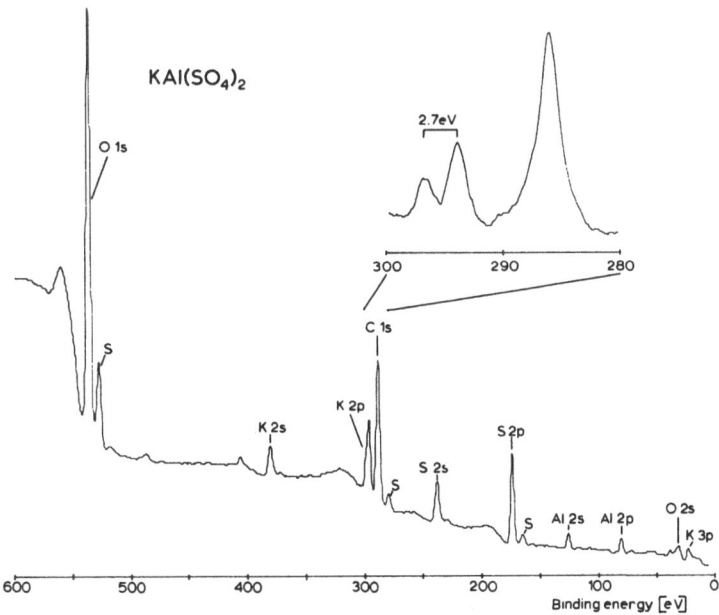

Fig. 9. Survey spectrum of 600 eV of KAl $(SO_4)_2$. The different features are clearly visible: photoelectron lines resulting from various orbitals, energy loss tails and X-ray satellites (S). The expanded insert of the potassium-carbon region reveals the characteristic spin orbit splitting of the K 2p-level. Mg Kα irradation, total observation time 30 min

17

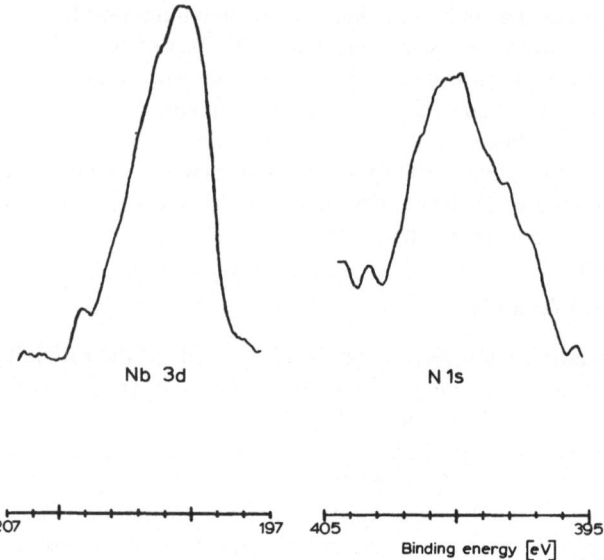

Fig. 10. Photoelectron lines of niobium and nitrogen. The spectra were obtained by sputtering a steel sample with a concentration of approximately 250 ppm NbN and by scanning the appropriate regions for 500 s each under high sensitivity conditions (E_A = 100 eV)

element, or by identifying in the spectrum the photoelectron lines of the other orbitals of the same element which should display a certain intensity relationship and approximately identical chemical shifts [17].

A question frequently asked is the detection limit of this technique. First, all elements of the periodic table can be observed with approximately equal sensitivity [23,62] except hydrogen which does not possess closed electron shells. However, XPS is by no means a trace method and concentrations below 100 ppm certainly cause problems. Fig. 10 exemplifies this with a steel sample containing 250 ppm of niobium and nitrogen. The spectra were obtained in an observation time of 500 sec each and show despite the short accumulation a good signal to noise ratio. On the other hand, if the element to be investigated is enriched on the surface, much lower concentrations than in the bulk material would be sufficient. Even fractions of a monolayer can be observed in the XPS-spectrum.

Three nice examples are worth mentioning: the moonrock spectra reported by L. Wilson [63], showing at the same time that iron is present as Fe (II). The end of the polywater hoax with the discovery that this mysterious compound consists mainly of sodiumsilicates, -sulfates and -lactates (from the transpiration of the hard working chemists!) [64]. And finally the air pollution studies of T. Novakow [65] working with very low quantities of smog particulates deposited on Teflon films and determining the lead concentration and the nitrogen and sulfur oxidation as function of the day time, thus allowing a much better insight into the kinetics of air pollution.

5.3 Quantitative Analysis

There is only a small step from the qualitative identification of a peak to the measurement of its integral. However, if a precise concentration determination is desired, certain precautions must be observed [17]. Because of the strong dependence of the signal intensity from the sample preparation, only concentrations relative to an internal standard can be measured. Either an element of high and constant concentration within the sample itself, or a homogeneous mixture with an added standard, e.g. LiF, can serve this purpose. The "relative photoelectric cross-sections" [1,17,179] should be calibrated or taken from reference tables [23,62]. The spectra of different spectral regions, i.e. of different elements, must be recorded with the sequential scanning technique in order to exclude time dependent intensity distortions due to surface contamination or destruction, and changes of the x-ray energy or the vacuum [17]. Before the actual digital integration the rising baseline of the photoelectron spectra (cf. Fig. 9) has to be corrected [17], possibly followed by a peak matching, if two closely spaced lines overlap like in the case of the $NaNO_2 / NaNO_3$-mixture [17].

The determination of the nitrogen content in different cereal grains represents a nice example [66,67]. The flour of the grain was mixed with 10% of lithium fluoride. The relative integrals F_N and F_F obtained are now plotted against the nitrogen content C_N determined by classical analytical techniques (Fig. 11). The bent slope of the curve can be easily explained even with a crude picture. The integral F is proportional to the intensity of the photoelectron beam I:

$$F = k \cdot I.$$

This intensity now depends on two factors: the concentration C_N of the nitrogen in the compound and the attenuation S by inelastic scattering:

$$I = k (1 - S) C_N.$$

In first approximation S should decrease with increasing nitrogen concentration:

$$S = k \cdot (1 - C_N)$$

thus leading to the final equation

$$I = k_1 C_N + k_2 C_N^2 \tag{7}$$

and under the assumption that I_F is constant (C_F = constant) to the formula inserted in Fig. 11.

19

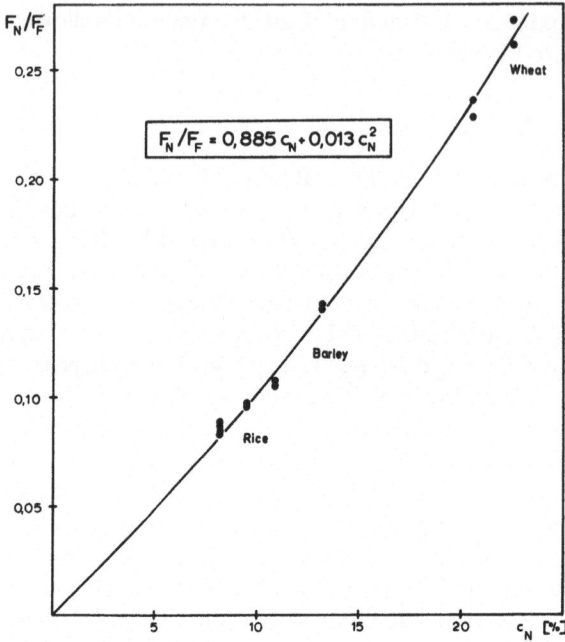

Fig. 11. Plot of relative integrals of the nitrogen photoelectron line (F_N/F_F) versus the concentration C_N of nitrogen in cereal grains [66]. The values are the mean of at least three integrations

Because the nitrogen concentration is a good measure for the total protein content of grain and thereby its nutritional value, the fast and relatively accurate XPS-measurements can help the plant breeder to improve the quantity of his grain protein. Other examples of quantitative analysis have been reported in the literature [1,79]. Besides the determination of bulk concentrations, quantitative studies on sample surfaces are easily possible and a sequence of measuring — sputtering — measuring — sputtering — etc. can help to determine concentration gradients of different elements near the surface.

5.4 Chemical Shift

The most important information in XPS originates from the small shifts of the binding energy induced by changes in the chemical environment of the atoms. For instance a positive charge, *i.e.* a lack of electrons, in the electron cloud creates an additional potential which the photoelectron has to overcome when it leaves the atom. Thus the binding energy E_B increases by ΔE_B which shows in the XPS-spectrum as a shift of the photoelectron line to lower

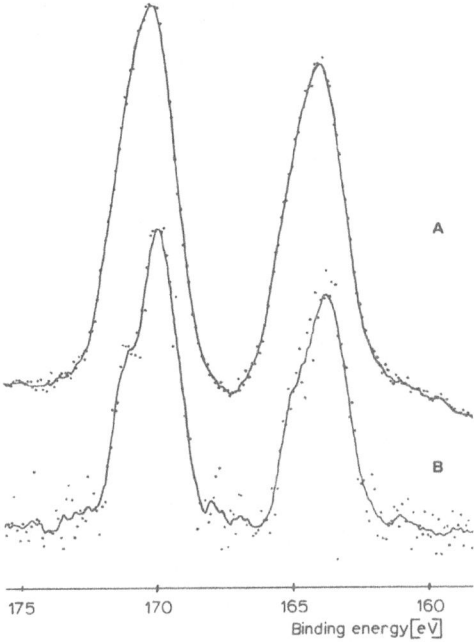

175   170   165   160
Binding energy[eV]

Fig. 12. Photoelectron spectra of sodiumthiosulfate $Na_2S_2O_3$ under different instrumental conditions. The sulphur peak at higher binding energies corresponds to the more positive central sulphur atom. Fig. 12 A is a high sensitivity scan (E_A = 100 eV) over a period of 100 s, while Fig. 12 B was run at higher resolution (E_A = 30 eV) and increased observation time (200 s). Note the spin orbit splitting of the sulphur 2p-doublet appearing in the second spectrum. The noise of the digitally filtered line is demonstrated by the accompanying point plot

kinetic energies. This phenomenon was called "chemical shift" [1] in analogy to a related parameter in NMR spectroscopy [68]. Fig. 12 illustrates this for a spectrum of sodiumthiosulfate which displays two lines for the positive sulphur (+6) on the left and the negative thio-sulphur (−2) on the right. Thiosulfate was actually the first evidence for a separation of two lines of the same element due to chemical shift effects [71].

There is quite a number of theoretical approaches to the understanding of experimentally observed chemical shifts. It was early realized that chemical shifts could be related to the formal oxidation state of the element under study. Further investigations revealed that the effective charge $q(A)$ of an atom A in a molecule is the important parameter and numerous correlations based on the equation

$$\Delta E_B = k\,q(A) \tag{8}$$

21

were published in the literature. However, many theoretical models with various levels of sophistication can be used. Simple correlations with the electronegativity of the neighbours [69] and with charges based on the concept of partial electronegativities [1,70] gave surprisingly good results. An example for 80 sulphur containing compounds [1] is depicted in Fig. 12. Similar relationships have been reported for boron [72], carbon [29,73], nitrogen [1,74,75], oxygen [29], silicon [76], phosphorous [77], sulphur [1,78], and chlorine [1].

In many cases more refined molecular orbital models give a better agreement between theory and experiment. Self-consistent field- [80,81] or CNDO- [82] calculations as well as other ab initio calculations [83,84] were performed and the results of several different approaches for phosphorous compounds were critically evaluated by M. Pelavin [77]. For the nitrogen compounds two different linear relationships, one for cations and for neutral molecules, the other for anoins, were observed [82], a phenomenon which might be explained with a Madelung contribution.

Even though correlations of binding energy with atomic charge are very useful in structure determination and studies of chemical bonds, one has to keep in mind that these models are approximations and the conclusions drawn are only valid within the specific limitations of each theoretical approach. Further sophisticated calculations [85,90] and correlations with ther-

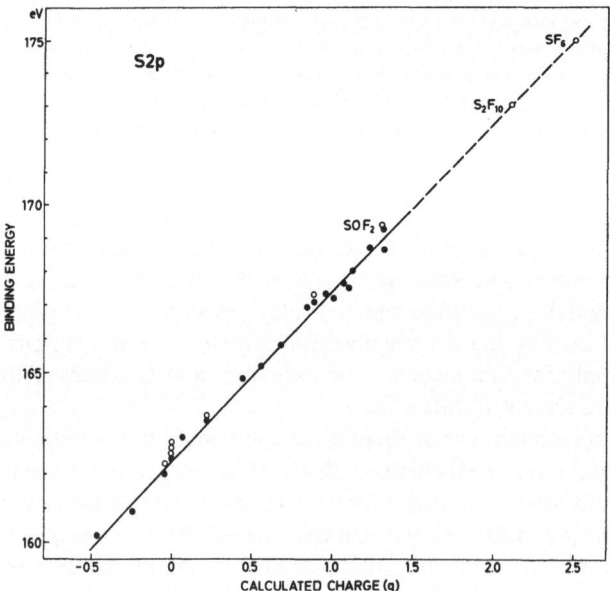

Fig. 13. Plot of the binding energies E_B of the sulphur $2p$-electrons versus calculated charge q for more than eighty compounds containing sulphur [1]

modynamic data [86] will certainly allow much better understanding of the theoretical background of the chemical shift, however, it does not influence the basic usefulness of first order approaches for the every day work of the analytical chemist.

5.5 Effects of the Crystal Lattice

C. Fadley and S. Hagström [87] were among the first to realize how important the contributions of the crystal lattice to the chemical shift of an ion are. This effect can be approximated with the Madelung-equation [88]:

$$\Delta E_B \sim \frac{q}{4\pi\epsilon} \left(\frac{1}{r_1} - \frac{1}{r_2} \right) \tag{9}$$

where ϵ is the dielectric constant and r_1 and r_2 the distances of the gegenions in the lattice.

This Madelung potential should be taken into account in all detailed studies of chemical shifts in ionic crystals. Examples of calculations [24,87,89,91] show the order of magnitude to be expected, but are still away from being accurate.

However, the practical application of this phenomenon for crystal structure investigations has recently been demonstrated by Freund [20]. There exists a linear relationship between the electron binding energies of different magnesium orbitals and the mean magnesium-oxygen distance as calculated from x-ray crystallographic data for a series of magnesium silicates. This relationship allowed an estimate of the structure of two x-ray amorphous compounds and confirmed predictions that a vacancy was formed in the crystal lattice.

There are certainly more effects to be considered, like polarisability of counterions [69,91] and all the questions connected with the persistance of atomic orbitals [92] and the applicability of Koopman's theorem [93,94]. However, as the article is mainly intended to reach the practically oriented chemist, the reader interested in theoretical considerations has to be referred to the literature.

5.6 Some Applications in Practical Chemistry

The applications of XPS-spectroscopy in practical chemistry are manifold and we can only touch some typical results in order to exemplify the potential of this technique. A more complete picture can be obtained by consulting the references listed in Table 2. As can be seen, nearly all elements of the periodic table have been investigated by XPS and the chemical shifts reported in the literature might help with the interpretation of data and can give a feeling, whether a problem is suitable for XPS-spectroscopy.

Table 2. Reference list for chemical shift data of most elements of the periodic table

Element		Authors	Year	Ref.
3	Lithium	Jørgensen	1972	95)
4	Beryllium	Jørgensen	1972	95)
		Hamrin, Siegbahn	1968	96)
5	Boron	Hendrickson, Hollander	1970	72)
		Bremser, Linnemann	1972	69)
6	Carbon	Siegbahn *et al.*	1969	29)
		Gelius, Siegbahn	1970	73)
		Hendrickson, Hollander	1970	72)
		Axelson, Siegbahn	1967	97)
		Olah, Mateescu	1970	98)
		Barber, Clark	1970	99)
		Schwartz, Coulson	1970	100)
		Thomas	1970	101)
		Mateescu, Riemenschneider	1971	102)
7	Nitrogen	Siegbahn *et al.*	1967	1)
		Siegbahn *et al.*	1969	29)
		Nordberg, Siegbahn	1968	74)
		Hendrickson, Hollander	1969	75)
		Finn, Hollander	1971	103)
		Jack, Hercules	1971	104)
		Patsch, Thieme	1971	105)
		Jørgensen	1972	95)
		Clark, Lilley	1971	106)
		Leigh, Bremser	1970	107)
8	Oxygen	Siegbahn *et al.*	1969	29)
		Yin, Adler	1971	108)
		Morgan, van Wazer	1971	109)
9	Fluorine	Hayes, Edelstein	1971	110)
		Jørgensen, Berthou	1972	89)
		Jørgensen	1972	95)
		Siegbahn *et al.*	1969	29)
11	Sodium	Jørgensen	1972	95)
12	Magnesium	Freund, Hamich	1971	20)
13	Aluminium	Ogilvie, Wolberg	1972	111)
14	Silicon	Nordberg, van Wazer	1970	76)
15	Phosphorous	Pelavin, Hollander	1970	77)
		Morgan, van Wazer	1971	109)
		Swartz, Hercules	1971	112)
		Bremser, Schipper	1971	17)
		Hedman, Klasson	1971	113)
		Blackburn, Nordberg	1970	114)
		Leigh, Bremser	1971	115)
		Jørgensen	1972	95)

Table 2 (continued)

Element		Authors	Year	Ref.
16	Sulphur	Siegbahn *et al.*	1967	1)
		Siegbahn *et al.*	1969	29)
		Lindberg, Siegbahn	1970	78)
		Lindberg, Hamrin	1970	116)
		Lindberg, Högberg	1971	117)
		Jørgensen	1972	95)
		Kramer, Klein	1969	118)
		Lindberg	1970	119)
		Clark, Lilley	1971	106)
		Patsch, Thieme	1971	105)
17	Chlorine	Fahlman, Siegbahn	1966	1)
		Jørgensen	1972	95)
19	Potassium	Jørgensen	1972	95)
		Jørgensen, Berthou	1972	89)
20	Calcium	Jørgensen	1972	95)
22	Titanium	Ramquist, Hamrin	1969	121)
		Jørgensen	1972	95)
23	Vanadium	Hamrin, Nordling	1970	122)
		Jørgensen	1972	95)
		Ramquist, Hamrin	1970	123)
24	Chromium	Hendrickson, Hollander	1970	72)
		Helmer	1970	124)
		Jørgensen	1972	95)
		Clark, Adams	1971	125,126)
25	Manganese	Jørgensen	1972	95)
26	Iron	Kramer, Klein	1969	118)
		Leibfritz, Bremser	1970	127)
		Winogradow, Nefedow	1971	128)
		Jørgensen	1972	95)
		Kramer, Klein	1971	130)
		Leibfritz	1972	129)
		Wertheim, Rosencwaig	1971	131)
		Clark, Adams	1971	125)
		Kramer, Klein	1971	132)
27	Cobalt	Jørgensen	1972	95)
		Bremser, Cardin	1971	17)
		Mavel, Escard	1971	133)
28	Nickel	Jørgensen	1971	24)
		Jørgensen	1972	95)
		Clark, Adams	1971	125)
29	Copper	Wolberg, Ogilvie	1970	134)
		Jørgensen	1972	95)
		Novakov, Prins	1971	135)

25

Table 2 (continued)

Element		Authors	Year	Ref.
30	Zinc	Mavel, Escard	1971	133)
		Jørgensen	1972	95)
		Langner, Vesely	1971	135,137)
31	Gallium	Jørgensen	1972	95)
33	Arsenic	Hulett, Carlson	1971	138)
		Jørgensen	1972	95)
34	Selenium	Malmsten, Högberg	1970	139)
		Swartz, Hercules	1971	140)
		Jørgensen	1972	95)
		Morgan, van Wazer	1971	109)
35	Bromine	Hulett, Carlson	1971	138)
		Jørgensen	1972	95)
37–40	Rb, Sr, Y, Zr	Jørgensen	1972	95)
41	Niobium	Ramquist, Nordling	1970	123)
		Jørgensen	1972	95)
42	Molybdenum	Swartz, Hercules	1971	79)
		Miller, Barber	1971	141)
		Jørgensen	1972	95)
44,46	Ru, Pd	Jørgensen	1972	95)
48	Cadmium	Vesely, Langner	1971	137)
		Jørgensen	1972	95)
50,51	Sn, Sb	Jørgensen	1972	95)
52	Tellurium	Swartz, Hercules	1971	140)
		Jørgensen	1972	95)
53	Jodine	Fadley, Hagström	1968	87,142)
		Jørgensen, Berthou	1972	89)
		Jørgensen	1972	95)
54	Xenon	Siegbahn et al.	1969	29)
55–57	Cs, Ba, La	Jørgensen	1972	95)
58–71	Rare earths	Fadley, Hagström	1968	87,142)
		Nilsson, Siegbahn	1968	143)
		Nilsson, Bergmark	1970	144)
		Jørgensen	1972	95)
		Hedén, Hagström	1971	55)
72–73	Hf, Ta	Jørgensen	1972	95)
74	Tungsten	Hamrin, Nordling	1970	122)
		Jørgensen	1972	95)
75	Rhenium	Leigh, Bremser	1971	115)
		Nefedow	1971	145)
		Jørgensen	1972	95)

Table 2 (continued)

Element		Authors	Year	Ref.
76, 77	Os, Ir	Nefedow	1971	[145]
		Jørgensen	1972	[95]
78	Platinum	Cook, Siegbahn	1971	[146]
		Riggs	1971	[147]
		Moddeman, Blackburn	1971	[148]
		Clark, Adams	1971	[149]
		Jørgensen	1972	[95]
79–83	Au, Hg, Tl, Pb, Bi	Jørgensen	1972	[95]
90, 92	Th, U	Jørgensen	1972	[95]
		Novakow	1969	[176]
95	Americium	Krause, Wuilleumier	1971	[150]

XPS has been applied extensively for the determination of molecular structure [50,151,152]. Following a reaction by measuring the spectra of the starting material and the reaction products reveals in many cases, which atoms were oxydised [28] or where a new bond was formed [151]. Symmetry considerations can help to distinguish between structures where the charge distribution is symmetrical or where a polarisation can be identified in the spectrum by the broadening, splitting, or even separation of peaks. For example, cystine dioxide *(I)* was assigned the thiolsulfonate structure because of the observed sulphurdoublet [1,153] and the oxydation product of 4,4'-diethylazobenzene could be interpreted as an azoxybenzene *(II)* [17].

In organic chemistry XPS spectra gave a direct evidence for a classical carbonium ion in *t*-butyl-hexafluoroantimonate *(III)* with a chemical shift

(I)   *(II)*   *(III)*

difference of 3.4 eV between the two carbon lines [98]. In the meantime the spectra of various organic ions have been recorded [102]. M. Patsch studied the charge distribution in mesoionic compounds *(IV)* structurally related to sydnone and showed that they exist predominantly in the betaine form [105]. A similar charge separation was found in phosphorous-ylides *(V)* clearly ruling out the existence of an ylen-structure [17].

(IV) (V) (VI)

A very interesting question in this context is the electron distribution in complexes involving the free electrons of a lone pair. The molecular nitrogen bound to rhenium in the phosphinecomplex *(VI)* is strongly polarized and the XPS-spectrum shows a shift of 2.1 eV between the two nitrogen atoms [107, 115]. In cobalt porphyrin-complexes of the vitamin B_{12}-type a strong π-back bonding could be detected increasing the negative charge on the central metal as compared with the cobalt salts of corresponding oxydation state [17]. Fig. 14 shows a series of iron compounds and reveals that the iron in the hexacyano-complexes, though formally of oxydation state $+2$ or $+3$, is more negative than the metallic iron indicated by the dotted line at 710 eV. Only in the prussiate $Na_2[Fe(CN)_5 NO]$ the complexed iron possesses approximately zero charge, because of the different π-acid strength of the nitrosyl-ligand, which is in good agreement with results obtained from Mössbauer-spectroscopy [154].

Besides information on the strong back donation effect in iron-complexes, Fig. 14 also reveals that the famous prussian-blue $KFe[Fe(CN)_6]$ contains two different irons with localized charge [127,131]. The sharp signal on the left at lowest binding energies corresponds to the $[Fe(CN)_6]^{4-}$ anion, while the broad peak at 713 eV can be assigned to the Fe^{3+} cation. The striking differences in line width observed in this series of iron spectra must be ascribed to the number of unpaired spins. While the zero- and low-spin complexes exhibit very narrow lines, the high spin irons appear as a rather broad hump. Effects of this kind will be considered in more detail in the last chapter.

An interesting study of the electron affinity of different ligands as well as comparisons between inner and outer ligands in iridium-complexes was reported by W. Nefedow [145] demonstrating clearly that XPS can furnish a lot of valuable information on transition metal complexes and organometallic compounds in general.

Many examples of this type are also found in biochemical molecules, an application which suffered so far under the requirement for very high sensitivity, but will be overcome with modern spectrometers. A series of ferredoxines and iron-proteides [129] as well as RNA and its nucleotide bases [155] were studied by observing the iron spectrum, but also sulphur becomes more and more interest-

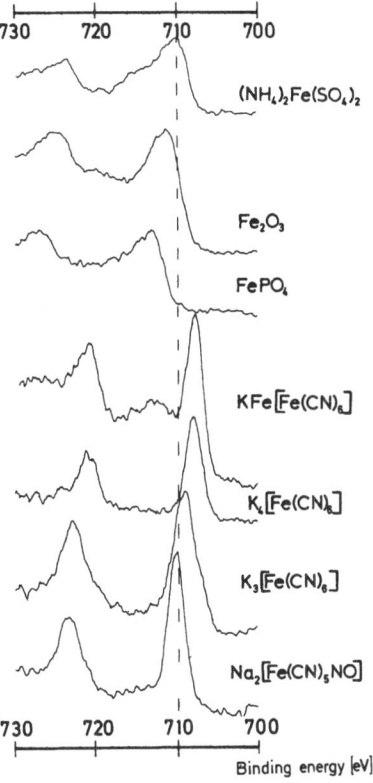

Fig. 14. Photoelectron spectra of the Fe 2p-spin doublet for a series of iron compounds [127]

ing [1]. The application of XPS to quantitative biological problems has already been mentioned above [66,67]. One important question in this connection is the investigation of frozen aqueous solutions [156], because many biological compounds cannot be isolated as solids and the technique of freeze drying [129] gives doubtful results.

Pollution studies is another area where XPS showed promising results. T. Novakov [65] demonstrated the potential of XPS for air pollution studies [157], while other investigations [138] detected arsenic in soil samples. Also the polywater controversy was revealed to be in reality a pollution problem [64].

The study of metals and metal surfaces is rapidly gaining general interest, mainly because of its importance in industrial research. Semiconductors [136] and especially the doping of semiconductors [158] are of prime interest. Studies of oxydation states were for instance carried out for W—V—O-phases [122] or for a series of carbides [123]. An example is shown in Fig. 10, where the state of the niobium and nitrogen present in very low concentrations in steel, had to be detected.

However, the most important application in this field is probably the study of metal surfaces in order to gain a deeper insight into the mechanisms of corrosion and catalysis. A review of some preliminary experiments can be found in the article of W. N. Delgass [159], including activation, aging and poisoning. Copper supported on alumina [134] as well as calcined and uncalcined $Mo-Al_2O_3$-systems [160] were investigated in detail. Especially the quantitative determination of molybdenum oxide mixtures [79] showed very promising results and could be applied also in catalysis studies. Surface oxidation of metallic magnesium [17] as well in platinum [161] pointed the way to studies of surface oxydation and corrosion. Generally speaking, comparisons of catalysts before and after the reaction [161], uncrushed and crushed to reveal the bulk properties, making use of the full potential of the reaction chamber for simulating reactions with gases under various conditions including poisoning, exposing deeper layers by sputtering, etc. will certainly yield a treasure of informations on the mechanism of catalysis and corrosion.

5.7 Some Theoretical Informations Contained in XPS-Spectra

The example of low- and high-spin iron in Fig. 14 revealed that the XPS spectra also contain information on the arrangement of unpaired spins in various shells of the atoms. C. S. Fadley was the first to observe core level splittings of up to 7 eV in Mn-, Fe-, Co- and Ni-3s orbitals due to the coupling of a hole in a metal-atom subshell to an unfilled valence subshell [168,169]. The core electrons with a spin parallel to the unpaired valence electron will experience an exchange potential reducing the average coulombic repulsion between two electrons. Thus parallel spins will be favoured energetically. A similar splitting was also reported for the Cr 3s-level [124] and the paramagnetic molecules oxygen (O_2) and nitrogenoxide (NO and NO_2) [29,163] (cf. Fig. 15). This phenomenon might help in the future to study organic free radicals [170] and to deduce a more detailed information on electron distribution and spin pairing.

Other nice examples of exchange splitting were reported for the 4f electrons of rare earth compounds [55,166], which show a single line for the half filled shell of Gd^{3+} and a second line of increasing intensity and decreasing distance for the heavier elements [166]. On the other hand, a different kind of splitting, called electrostatic splitting, was observed by T. Novakov [176] in the spectra of heavy elements (Th, U). In contrast to the d- and f-levels, the $5p_{3/2}$ lines are structured because of the differential interaction of the internal electrostatic field with the substates of the 5p electron. Theoretical calculations for gold compounds [177] pointed out, however, that the electrons of the ligands must be considered as well.

Configuration interaction was shown to be responsible for satellite lines in XPS-spectra of alkali halides [167]. If there are other final states present in the atomic system with the same configuration, i.e. same angular momentum

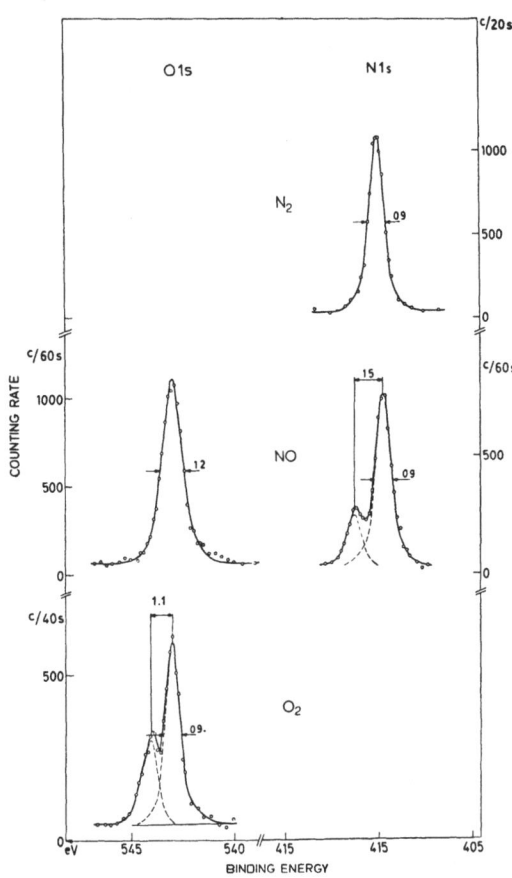

Fig. 15. Core level splitting of the 1s-photoelectron lines in the paramagnetic molecules NO and O_2 [29]. While molecular nitrogen N_2 with a closed shell configuration shows no splitting, the NO^+-ion formed in the photoelectron process can exist in a singlet or a triplet state giving rise to the observed N 1s-doublet with an intensity ratio of 1 : 3, while the oxygen line is only broadened. Molecular oxygen O_2 finally displays a splitting of 1.1 eV and a ratio of 1 : 2. All spectra were recorded in the gas phase [29]

and parity, but somewhat greater binding energy than the single-hole final states, perturbation theory predicts the appearance of satellites which become less and less in intensity the further they are shifted from the parent peak [167]. The multiple signals in the region of the 3d-levels of lanthanum(III) salts have been explained by Jørgensen [180] as a simultaneous transfer of an electron from the neighbour atoms to the empty 4f-shell during the ionisation.

This phenomenon is closely related to the monopole-excitation [171] or shake-up process [152] observed for several atoms and molecules. Noble gases

[29,171,172] were most extensively studied and the results compared with theoretical calculations. The intensity of these shake-up peaks can be relatively high, sometimes more than 20% of the main line. An interesting example is presented in the spectrum of carbonsuboxide (C_3O_2) [173]. From the intensity distribution U. Gelius [173] concluded that there is no C 1s shake-up transition associated with the central (more negative) carbon corresponding to the first O 1s shake-up line of C_3O_2, while there is a characteristic satellite pattern connected with the other carbon atoms.

Similar effects have also been observed for solids and seem to become a very important source of information, because for a given ion the satellites are found to depend in intensity, position and shape on the type of ligand in transition metal complexes [24,135,164,174,175]. Thus in many cases the study of shake-up peaks has provided additional evidence for a specific oxydation state [17] and will certainly increase our understanding of crystal and ligand field effects.

Finally one parameter in XPS, which so far has been neglected to a great deal, should be mentioned briefly: the angular distribution of the emitted photoelectrons. Studies on sodium chloride single crystals [165,178] showed promising results and might open a new field of application in crystal structure determinations.

6. Conclusion

The numerous applications in various fields of chemistry and physics have clearly demonstrated the potential of X-ray photoelectron spectroscopy. A number of interesting experimental projects will be completed in the near future and papers with new and more reliable reference data will appear in the literature. Further refinement of the theoretical models will add to the fundamental understanding of the obtained results. And last, but not least new developments in instrumentation will keep pace with the practical and theoretical experience and open up new areas, which so far could not be penetrated because of resolution, sensitivity or sample handling problems.

Acknowledgement. The author is indebted to F. Linnemann for his constant valuable help and numerous authors for relating their results prior to publication.

Literature

[1] Siegbahn, K., *et al.*: ESCA. Uppsala: Almquist & Wiksells Boktryckeri AB 1967.
[2] Robinson, H.: Proc. Roy. Soc. *A 104*, 455 (1923).
[3] de Broglie, M.: C. R. hebd. Séances Acad. Sci. *172*, 274 (1921).

[4] Siegbahn, K.: Alpha-, beta- and gamma-ray spectroscopy. Amsterdam: North-Holland Publ. Co. 1965. – Siegbahn, K., Edvarson, K.: Nucl. Phys. *1*, 137 (1956).
[5] Nordling, C., Sokolowski, E., Siegbahn, K.: Arkiv Fysik *13*, 483 (1958). – Sokolowski, E., Nordling, C., Siegbahn, K.: Phys. Rev. *110*, 776 (1958).
[6] Hagström, S., Nordling, C., Siegbahn, K.: Phys. Letters *9*, 235 (1964).
[7] Helmer, J. C., Weichert, N. H.: Appl. Phys. Letters *13*, 266 (1968).
[8] Weichert, N. H., Helmer, J. C.: Advances in x-ray analysis, Vol. 13, p. 406. New York: Plenum Press 1970.
[9] Siegbahn, K., Edvarson, K.: Nucl. Phys. *1*, 137 (1956). – Nordberg, R., Brecht, H., Albridge, R., Fahlman, A., van Wazer, J.: J. Inorg. Chem. *9*, 2469 (1970).
[10] Fahlman, A., Hagström, S., Hamrin, K., Nordberg, R., Nordling, C., Siegbahn, K.: Arkiv Fysik *31*, 479 (1966).
[11] Bremser, W.: Z. Anal. Chem. *259*, 204 (1972).
[12] Brundle, C. R., Robin, M. B., in: Determination of organic structures by physical methods, Vol. 3 (ed. F. C. Nachod and J. J. Zuckerman). New York: Academic Press 1971.
[13] Turner, D. W.: Molecular photoelectron spectroscopy. In: Physical methods in advanced inorganic chemistry. Interscience Publ. 1968.
[14] Heilbronner, E., Muszkat, K. A., Schäublin, J.: Helv. Chim. Acta *54*, 58 (1971).
[15] Comes, F. J.: Messtechnik *77*, 87 (1969).
[16] Turner, D. W., Baker, C., Baker, A. D., Brundle, C. R.: Molecular photoelectron spectroscopy. New York: Wiley Interscience Publ. 1970. – Baker, A. D.: Accounts Chem. Res. *3*, 17 (1970).
[17] Bremser, W.: Chemiker-Ztg. *95*, 819 (1971).
[18] Hercules, D. M.: Anal. Chem. *42*, 20A (1970).
[19] Hagedorn, H. L., Wapstra, A. H.: Nucl. Phys. *15*, 146 (1960).
[20] Freund, F., Hamich, M.: Fortschr. Mineral. *48*, 243 (1971).
[21] Harris, L. A.: Anal. Chem. *40*, 24A (1968).
[22] Harris, L. A.: J. Appl. Phys. *39*, 1419 (1968). – Palmberg, P. W., Rhodin, T. N.: J. Appl. Phys. *39*, 2425 (1968).
[23] Wagner, C. D.: Anal. Chem. *44*, 1050 (1972).
[24] Jørgensen, C. K.: Chimia *25*, 213 (1971).
[25] Wagner, C. D.: Anal. Chem. *44*, 967 (1972). – Ref. [26], Chap. VII. 1.
[26] Conference abstracts of the International Conference on Electron Spectroscopy, Asilomar (Calif.), September 1971.
[27] Bodmer, A.: The photoelectron spectrometer. In: Conference abstracts "X-ray photoelectron spectroscopy", p. 1. Zürich, October 1971.
[28] Bremser, W.: Messtechnik *78*, 133 (1970).
[29] Siegbahn, K., *et al.*: ESCA applied to free molecules. Amsterdam: North-Holland Publ. Co. 1969.
[30] Siegbahn, K.: Perspectives and problems in electron spectroscopy, Uppsala University Institute of Physics, Publication Nr. 752, October 1971.
[31] Brügel, W., Holm, R.: Private communication.
[32] Pines, D.: Elementary excitations in solids. New York: Benjamin Inc. 1963.
[33] Larsson, K., Nordling, C., Siegbahn, K., Stenhagen, E.: Acta Chem. Scand. *20*, 2880 (1966).
[34] Baer, Y., Hedén, P. F., Hedman, J., Klasson, M., Nordling, C.: Solid State Commun. *8*, 1479 (1970).
[35] Sokolowski, E.: Arkiv. Fysik *15*, 1 (1959).
[36] Hnatowich, D. J., Hudis, J., Perlman, M. L., Ragaini, R. C.: J. Appl. Phys. *42*, 4883 (1971).

W. Bremser

37) Bremser, W., Linnemann, F.: Chemiker-Ztg. 95, 1011 (1971).
38) Brundle, C. R.: Appl. Spectry. 25, 8 (1971).
39) Fadley, C. S., Hagström, S. B. M., Klein,.M. P., Shirley, D. A.: J. Chem. Phys. 48, 3779 (1968).
40) Siegbahn, K.: Phil. Trans. Roy. Soc. London A 268, 33 (1970).
41) Nordling, C., Siegbahn, K.: Rev. Roumaine Phys. 11, 797 (1966).
42) Shirley, D.: Science 161, 745 (1968).
43) Berry, R. S.: Ann. Rev. Phys. Chem. 20, 357 (1969).
44) Hercules, D. M.: Anal. Chem., in press.
45) Jørgensen, C. K.: Modern aspects of ligand field theory. Amsterdam: North-Holland Press 1971.
46) Hollander, J. M., Jolly, W. L.: Accounts Chem. Res. 3, 193 (1970).
47) Nordberg, R.: Advan. X-Ray Anal. 13, 390 (1970).
48) Pignataro, S.: Chim. Ind. 53, 382 (1971).
49) Mavel, G.: Bull. Soc. Chim. France 1970, 3303.
50) Lindberg, B.: XXIII IU PAC Congress Boston 1971. – Lindberg, B., in: Molecular Spectroscopy 1971, p. 61. London 1971.
51) Nordling, C.: Angew. Chem. 84, 144 (1972).
52) Hagström, S. B. M.: Photoelectron spectroscopy to study the electronic band structure of solids. In: Conference abstracts "X-ray photoelectron spectroscopy". Zürich, October 1971.
53) Baer, Y., Hedén, P. O., Hedman, J., Klasson, M., Nordling, C., Siegbahn, K.: Phys. Scr. 1, 55 (1970).
54) Baer, Y., Hedén, P. O., Hedman, J., Klasson, M., Nordling, C., Siegbahn, K.: Solid State Commun. 8, 517 (1970).
55) Hedén, P. F., Löfgren, H., Hagström, S. B. M.: Phys. Rev. Letters 26, 432 (1971).
56) Hagström, S. B. M., Hedén, P. F., Löfgren, H.: Solid State Commun. 8, 1245 (1970).
57) Gelius, U., Allan, C. J., Johannson, G., Siegbahn, H., Allison, D. A., Siegbahn, K.: The ESCA spectra of benzene and the iso-electronic series, thiophene, pyrrole and furan. Uppsala University Institute of Physics, Publication Nr. 746, August 1971.
58) Thomas, J., Evans, E., Barber, M., Swift, P.: Trans. Faraday Soc. 67, 1875 (1971).
59) Hamrin, K., Johansson, G., Gelius, U., Nordling, C., Siegbahn, K.: Phys. Scr. 1, 277 (1970).
60) Prins, R., Novakov, T.: Chem. Phys. Lett. 9, 593 (1971).
61) Shirley, D. A.: Ref. 26), Chap. IV. 10.
62) Berthou, H., Jørgensen, C. K.: To be published.
63) Wilson, L. A., Manatt, S., Huntress, W. T.: Unpublished spectrum shown in Ref. 17).
64) Davis, R. E., Rousseau, D. L., Board, R. D.: Science 171, 167 (1971).
65) Novakov, T.: Am. Inst. of Aeronautics and Astronautics, paper no. 71–1103 (1971).
66) Bremser, W., Luse, R. A., Linnemann, F.: Unpublished.
67) Klein, M., Kramer, L.: Proceedings of the Symposium on Improving Plant Protein by Nuclear Techniques.
68) Proctor, W. G., Yu, F. C.: Phys. Rev. 77, 717 (1950).
69) Bremser, W., Linnemann, F.: Chemiker-Ztg. 96, 36 (1972).
70) Fahlmann, A., Hamrin, K., Hedman, J., Nordberg, R., Nordling, C., Siegbahn, K.: Nature 210, 4 (1966).
71) Hagström, S. B. M., Nordling, C., Siegbahn, K.: Z. Physik 178, 439 (1964).
72) Hendrickson, D. N., Hollander, J. M., Jolly, W. L.: Inorg. Chem. 9, 612 (1970).
73) Gelius, U., Hedén, P. F., Hedman, J., Lindberg, B. J., Manne, R., Nordberg, R., Nordling, C., Siegbahn, K.: Phys. Scr. 2, 70 (1970).
74) Nordberg, R., Albridge, R. G., Bergmark, T., Eriksson, U., Hedman, J., Nordling, C.,

Siegbahn, K., Lindberg, B. L.: Arkiv Kemi *28*, 257 (1968).

[75] Hendrickson, D. N., Hollander, J. M., Jolly, W. L.: Inorg. Chem. *8*, 2642 (1969).

[76] Nordberg, R., Brecht, H., Albridge, R. C., Fahlman, A., van Wazer, J. R.: Inorg. Chem. *9*, 2469 (1970).

[77] Pelavin, M., Hendrickson, D. N., Hollander, J. M., Jolly, W. L.: J. Phys. Chem. *74*, 1116 (1970).

[78] Lindberg, B. J., Hamrin, K., Johansson, G., Gelius, U., Fahlman, A., Nordling, C., Siegbahn, K.: Phys. Scr. *1*, 286 (1970).

[79] Swartz, E., Hercules, D. M.: Anal. Chem. *43*, 1774 (1971).

[80] Basch, H., Snyder, L. C.: Chem. Phys. Lett. *3*, 333 (1969).

[81] Wyatt, J., Hillier, I., Saunders, V., Connor, J., Barber, M.: J. Chem. Phys. *54*, 5311 (1971).

[82] Hollander, J., Hendrickson, D., Jolly, W. L.: J. Chem. Phys. *49*, 3315 (1968).

[83] Gelius, U., Roos, B., Siegbahn, P.: Chem. Phys. Lett. *4*, 471 (1970).

[84] Schwartz, M.: Chem. Phys. Lett. *7*, 78 (1970).

[85] Barber, M., Clark, D. T.: Chem. Commun. *1970*, 22.

[86] Jolly, W. L.: J. Am. Chem. Soc. *92*, 3260 (1970).

[87] Fadley, C. S., Hagström, S. B. M., Klein, M. P., Shirley, D. A.: J. Chem. Phys. *48*, 3779 (1968).

[88] Madelung, E.: Z. Physik *19*, 528 (1918).

[89] Jørgensen, C. K., Berthou, H., Balsenc, L.: J. Fluorine Chem. *1*, 327 (1971).

[90] Hohlneicher, G., Ecker, F., Cederbaum, L.: Ref. [26], Chap. V. 1.

[91] Critrin, P. H., Shaw, R. W., Packer, A., Thomas, T. D.: Ref. [26], Chap. V. 4.

[92] Jørgensen, C. K.: Accounts Chem. Res. *4*, 307 (1971).

[93] Manne, R., Aberg, T.: Chem. Phys. Lett. *7*, 282 (1970).

[94] Snyder, L.: J. Chem. Phys. *55*, 95 (1971). – Davis, D. W., Hollander, J. M., Shirley, D. A., Thomas, T. D.: J. Chem. Phys. *52*, 3295 (1970).

[95] Jørgensen, C. K.: Theoret. Chim. Acta *24*, 241 (1972). – Jørgensen, C. K., Berthou, H.: Mat. Fys. Medd. Danske Vid. Selskab: In press.

[96] Hamrin, K., Johansson, G., Fahlman, A., Nordling, C., Siegbahn, K.: Nucl. Sci. Abstr. *22*, 13211 (1968).

[97] Axelson, G., Ericson, N., Fahlman, A., Hamrin, K., Hedman, J., Nordberg, R., Nordling, C., Siegbahn, K.: Nature *213*, 70 (1967).

[98] Olah, G. A., Mateescu, G. D., Wilson, L. A., Gross, M. H.: J. Am. Chem. Soc. *92*, 7231 (1970).

[99] Barber, M., Clark, D. T.: Chem. Commun. *1970*, 23, 24.

[100] Schwartz, M., Coulson, C., Allen, L.: J. Am. Chem. Soc. *92*, 447 (1970).

[101] Thomas, T.: J. Am. Chem. Soc. *92*, 4184 (1970).

[102] Mateescu, G. D., Riemenschneider, J. L.: Ref. [26], Chap. V. 17.

[103] Finn, P., Pearson, R. K., Hollander, J. M., Jolly, W. L.: Inorg. Chem. *10*, 378 (1971).

[104] Jack, J. J., Hercules, D. M.: Anal. Chem. *43*, 729 (1971).

[105] Patsch, M., Thieme, P.: Angew. Chem., Int. Ed. *10*, 569 (1971).

[106] Clark, D. T., Lilley, D. M. J.: Chem. Phys. Lett. *9*, 234 (1971).

[107] Leigh, G. J., Murrell, J. N., Bremser, W., Proctor, W. G.: Chem. Commun. *1970*, 1661.

[108] Yin, L., Ghose, S., Adler, I.: Science *173*, 633 (1971).

[109] Morgan, W., Stec, W., Albridge, R., van Wazer, J.: Inorg. Chem. *10*, 926 (1971).

[110] Hayes, R. G., Edelstein, N.: Ref. [26], Chap. V. 16.

[111] Ogilvie, J. L., Wolberg, A.: J. Appl. Spectr., submitted for publ.

[112] Swartz, W. E., Hercules, D. M.: Anal. Chem. *43*, 1066 (1971).

[113] Hedman, J., Klasson, M., Nordling, C., Lindberg, B. J.: Ref. [26], Chap. V. 9.

[114] Blackburn, J., Nordberg, R., Stevie, F., Albridge, R. G., Jones, M. M.: Inorg. Chem.

W. Bremser

9, 2374 (1970).
[115] Leigh, G. J., Bremser, W., in: Molecular Spectroscopy 1971, p. 93. London 1971.
[116] Lindberg, B. J., Hamrin, K.: Acta Chem. Skand. 24, 3661 (1970).
[117] Lindberg, B. J., Högberg, S., et al.: Chem. Scr. 1, 183 (1971). – Gleiter, R., Hornung, V., Lindberg, B., Högberg, S., Lozac'h, N.: Chem. Phys. Lett. 11, 401 (1971).
[118] Kramer, L., Klein, M.: J. Chem. Phys. 51, 3618 (1969).
[119] Lindberg, B. J.: Acta Chem. Scand. 24, 2242 (1970).
[120] Fahlmann, A., Carlsson, R., Siegbahn, K.: Arkiv Kemi 25, 301 (1966).
[121] Ramquist, L., Hamrin, K., Johansson, G., Fahlman, A., Nordling, C.: J. Phys. Chem. Solids 30, 1835 (1969); – cf. Ref. [123].
[122] Hamrin, K., Nordling, C., Kihlborg, L.: Ann. Acad. Reg. Sci. Upsalien. 14, 1 (1970).
[123] Ramquist, L., Hamrin, K., Johansson, G., Gelius, U., Nordling, C.: J. Chem. Phys. Solids 31, 2669 (1970).
[124] Helmer, J. C.: Unpublished, presented at the Conference on Photoelectron Spectroscopy, Oxford, September 1970.
[125] Clark, D. T., Adams, D. B.: J. Chem. Soc. D 1971, 740.
[126] Clark, D. T., Adams, D. B.: Chem. Phys. Lett. 10, 121 (1971).
[127] Leibfritz, D., Bremser, W.: Chemiker-Ztg. 94, 882 (1970).
[128] Winogradow, A. P., Nefedow, W. I., Urusow, W. S., Schaworonkow, N. M.: Dokl. Nauk SSR 201, 957 (1971).
[129] Leibfritz, D.: Angew. Chem. 84, 156 (1972).
[130] Kramer, L., Klein, M.: Chem. Phys. Lett. 8, 183 (1971).
[131] Wertheim, G., Rosencwaig, A.: J. Chem. Phys. 54, 3235 (1971).
[132] Kramer, L. N., Klein, M. P.: Ref. [26], Chap. V. 11.
[133] Escard, J., Mavel, G.: ESCA – structural applications in the field of metal complexes, in: Conference abstracts "X-ray photoelectron spectroscopy", p. 47. Zürich, October 1971.
[134] Wolberg, A., Ogilvie, J. L., Roth, J. F.: J. Catalysis. 19, 86 (1970).
[135] Novakov, T.: Phys. Rev. B 3, 2693 (1971). – Novakov, T., Prins, R.: Ref. [26], Chap. IV. 8.
[136] Langner, D. W., Vesely, C. J.: Internal report, to be published in Phys. Rev. B.
[137] Vesely, C. J., Langner, D. W., Hengehold, R. L.: Ref. [26], Chap. IV. 4.
[138] Hulett, L. D., Carlson, T. A.: Appl. Spectry. 25, 33 (1971).
[139] Malmsten, G., Thorén, I., Högberg, S., Bergmark, J.-E., Karlsson, S. E.: Uppsala University Institute of Physics Report Nr. 682, April 1970; – Phys. Scr. 3, 96 (1971).
[140] Swartz, W. E., Wynne, K. J., Hercules, D. M.: Anal. Chem. 43, 1884 (1971).
[141] Miller, A., Atkinson, W., Barber, M., Swift, P.: J. Catalysis. 22, 140 (1971).
[142] Fadley, C. S., Hagström, S. B. M., Hollander, J. M., Klein, M. P., Shirley, D. A.: Science 157, 1571 (1967).
[143] Nilsson, O., Nordberg, C., Bergmark, J., Fahlman, A., Nordling, C., Siegbahn, K.: Helv. Phys. Acta 41, 1064 (1968).
[144] Nilsson, O., Bergmark, J., Thorén, I.: Uppsala Univ., Nucl. Sci. Abstr. 24, 21550 (1970); – 24, 21551 (1970).
[145] Nefedow, W. I.: Investigation of Pt-, Ir- and Re-complexes by means of ESCA. In: Conference abstracts "X-ray photoelectron spectroscopy", p. 57. Zürich, October 1971.
[146] Cook, C. D., Wan, K. Y., Gelius, U., Hamrin, K., Johansson, G., Olsson, E., Siegbahn, H., Nordling, C., Siegbahn, K.: J. Am. Chem. Soc. 93, 1904 (1971).
[147] Riggs, W. M.: Ref. [26], Chap. V. 2.
[148] Moddeman, W. E., Blackburn, J. R., Kumar, G., Morgan, K. A., Jones, M. M., Albridge, R. G.: Ref. [26], Chap. V. 7.
[149] Clark, D. T., Adams, D. B., Briggs, D.: Chem. Commun. 1971, 602.

150) Krause, M. O., Wuilleumier, F.: Ref. 26), Chap. V. 14.
151) Hedman, J., Hedén, P.-F., Nordberg, R., Nordling, C., Lindberg, B. J.: Spectrochim. Acta 26A, 761 (1970).
152) Allan, C. J., Siegbahn, K.: Electron spectroscopy for chemical application. Uppsala University Institute of Physics Publication Nr. 754, November 1971.
153) Axelson, G., Hamrin, K., Fahlman, A., Nordling, C., Lindberg, B. J.: Spectrochim. Acta 23A, 2015 (1967).
154) Fluck, E.: Fortschr. Chem. Forsch. 5, 395 (1966).
155) Hulett, L., Carlson, T.: Clin. Chem. 16, 677 (1970).
156) Kramer, L., Klein, M.: J. Chem. Phys. 51, 3620 (1969).
157) Araktingi, Y. E., Bhacca, N. S., Proctor, W. G., Robinson, J .W.: To be published.
158) Sharma, J., Staley, R., Rimstidt, J., Fair, H., Gora, T.: Chem. Phys. Lett. 9, 564 (1971).
159) Delgass, W. N., Hughes, T. R., Fadley, C. S.: Cat. Rev. 4, 179 (1970).
160) Miller, A., Atkinson, W., Barber, M., Swift, P.: J. Catalysis 22, 140 (1971).
161) Kim, K. S., Winograd, N., Davis, R. E.: J. Am. Chem. Soc. 93, 6296 (1971).
162) Delgass, W. N.: Unpublished, described in Ref. 46).
163) Hedman, J., Hedén, P. F., Nordling, C., Siegbahn, K.: Phys. Lett. 29A, 178 (1969).
164) Pignataro, S.: Z. f. Naturforschung, in press.
165) Siegbahn, K., Gelius, U., Siegbahn, H., Olson, E.: Phys. Lett. 32A, 221 (1970).
166) Wertheim, G. K., Rosencwaig, A., Cohen, R. L., Guggenheim, H. J.: Phys. Rev. Letters 27, 505 (1971).
167) Wertheim, G. K., Rosencwaig, A.: Phys. Rev. Letters 26, 1179 (1971).
168) Fadley, C. S., Shirley, D. A.: Phys. Rev. A2, 1109 (1970).
169) Fadley, C. S., Shirley, D. A., Freeman, A. J., Bagus, P. S., Mallow, J. V.: Phys. Rev. Letters 23, 1397 (1969).
170) Lindberg, B. J., Lemaire, H.: Reported in Ref. 50b).
171) Carlson, T. A., Krause, M. O., Moddeman, W. E.: J. Phys. C2, 102 (1971); – Phys. Rev. A1, 1406 (1970).
172) Krause, M. O., Carlson, T. A., Dismukes, R. D.: Phys. Rev. 170, 37 (1968).
173) Gelius, U., Allan, C. J., Allison, D., Siegbahn, H., Siegbahn, K.: Chem. Phys. Lett. 11, 224 (1971).
174) Barber, M., Connor, J. A., Hillier, I. H.: Chem. Phys. Lett. 9, 570 (1971).
175) Rosencwaig, A., Wertheim, G. K., Guggenheim, H. J.: Phys. Rev. Letters 27, 479 (1971).
176) Novakov, T., Hollander, J. M.: Phys. Rev. Letters 21, 1133 (1969).
177) Apai, G., Hollander, J. M., Novakov, T., Schaeffer, F., Shirley, D. A.: Unpublished, reported in Ref. 46).
178) Siegbahn, K., Gelius, U., Siegbahn, H., Olson, E.: Phys. Scr. 1, 272 (1970).
179) Gelius, U.: Molecular orbitals and line intensities in ESCA spectra. Uppsala University Institute of Physics, Publication Nr. 753, November 1971.
180) Jørgensen, C. K., Berthou, H.: Chem. Phys. Lett. 13, 186 (1972).

Received April 13, 1972

Synthesis of Organic Compounds in Glow and Corona Discharges

Prof. Dr. Harald Suhr

Chemisches Institut der Universität Tübingen, Tübingen

Contents

I. Introduction

Many attempts have been made to effect chemical changes by means of electrical discharges. As early as 1796 the conversion of ethylene to an oily substance under the influence of sparks was reported [1]. Towards the end of the last century numerous compounds were subjected to corona discharges [2]. When high-voltage, shortwave and microwave discharges became available they, too, were tested for their influence on various chemicals. The results of these experiments, which are reported in several hundred papers, can be summarized as follows:

1. Various forms of electrical discharges can effect chemical changes.

2. Many inorganic and practically all organic compounds can be destroyed by electrical discharges.

3. Often products with higher molecular weights than the starting material, or polymers, are formed.

4. Some experiments yielded products which could have been formed only by some drastic changes within the molecules or by unusual mechanisms.

The significance of these observations for preparative chemistry was soon realized. However, attempts to apply discharge techniques in chemical synthesis had only limited success.

Of all the reactions studied, only the synthesis of nitrogen oxides and acetylene in arcs or plasma torches and that of ozone in glow and corona discharges are of major importance. In addition, a few small-scale preparations of inorganic compounds have been developed, *e.g.* synthesis of hydrazine and of hydrides and halides of silicon, germanium, tin, lead, phosphorus or arsenic [3]:

$$2\,NH_3 \;\to\; H_2N-NH_2 + H_2$$

$$SiH_4 \;\to\; Si_2H_6, Si_3H_8, Si_4H_{10} + \ldots$$

$$2\,PCl_3 \;\to\; P_2Cl_4 + Cl_2$$

When organic compounds are subjected to the extreme temperatures present in arcs or plasma torches they are completely pyrolysed. Equally "cold" glow and corona discharges with low gas temperatures can destroy these compounds. In all early attempts to synthesize organic compounds in glow or corona discharges yields were small and most of the starting material was converted to tars and polymers. These experiments, although of little value to preparative chemistry, provided interesting information on chemical evolution [4]. Mixtures of simple molecules like methane, ammonia and water, which are suitable models for the atmosphere of the primitive earth, can produce amino acids under the influence of glow or corona discharges [5].

The synthesis of organic compounds on the preparative scale has been achieved, only very recently. Typical conditions are low power levels in the glow or corona discharge and fast flow rates for the organic material. Fig. 1

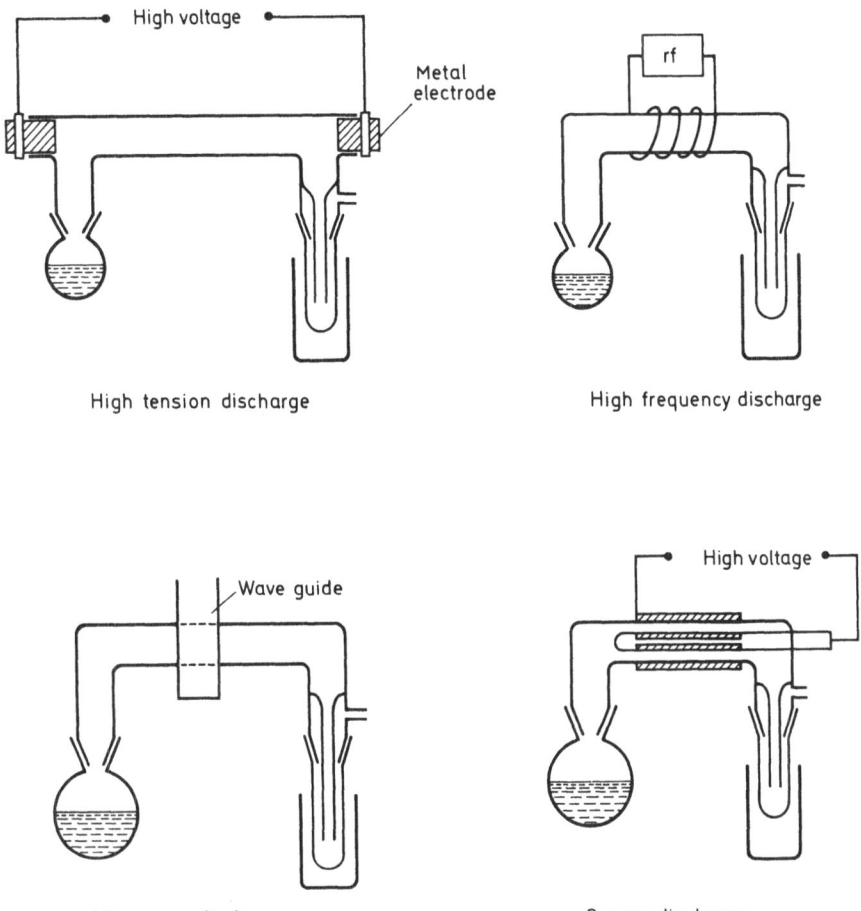

Fig. 1. Typical arrangements of organic plasma reactions

demonstrates several possible set-ups for such experiments. In d.c. or low-frequency glow discharges a high voltage is fed to metal electrodes inside a reaction vessel. In high-frequency arrangements a rf generator is coupled to the plasma by means of a coil or two metal rings encircling the reaction tube. For microwave glow discharges the reaction tube passes through a resonance cavity. In corona discharges a high voltage is first passed through a glass wall, then the vapor of the organic material, then another glass wall.

The results of plasma reactions are strongly dependent on pressure, electrical field strength, and gas velocity. Most experiments have been carried out at pressures of 1–5 Torr, power levels of approximately 100 W, and flow

41

rates of several meters per second. In a typical run, about 50 g of an organic substance pass through a discharge zone in one hour and some 30 per cent of this material is converted to products.

II. Applications

A. Reactions of Atomic Gases with Organic Molecules

A very simple example of organic plasma chemistry is the reaction of atomic gases with organic molecules. Normally atoms are generated through glow discharges and mixed with the organic material outside the discharge zone. Occasionally, higher yields have been obtained when mixtures of the molecular gases and the organic substances pass through the discharge zone together.

The atoms of hydrogen, oxygen, nitrogen and halogen react readily with various organic molecules [6]. Atomic hydrogen abstracts hydrogen or other atoms from saturated organic compounds. Unsaturated compounds add hydrogen atoms. In both cases radicals are formed.

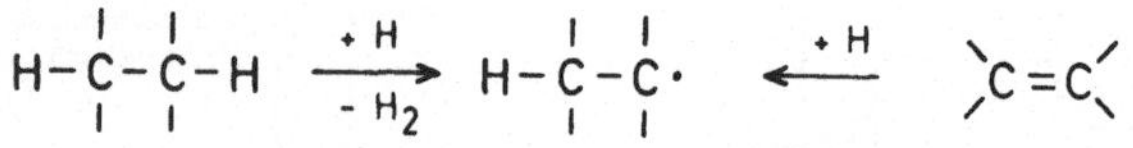

$$H-\overset{|}{\underset{|}{C}}-\overset{|}{\underset{|}{C}}-H \xrightarrow[-H_2]{+H} H-\overset{|}{\underset{|}{C}}-\overset{|}{\underset{|}{C}}\cdot \xleftarrow{+H} \enspace \overset{\diagdown}{\diagup}C=C\overset{\diagup}{\diagdown}$$

In the reaction with atomic nitrogen practically all organic compounds are destroyed. The products are mainly hydrogen cyanide, some cyanide and ammonia, low-molecular-weight gases and polymers [7]. Very few cases are known where nitrogen is incorporated into the products. When butadiene is allowed to react with atomic nitrogen in addition to hydrogen cyanide some 20 per cent of the yield consists of such nitrogen compounds as pyrrole and unsaturated nitriles [8-9]:

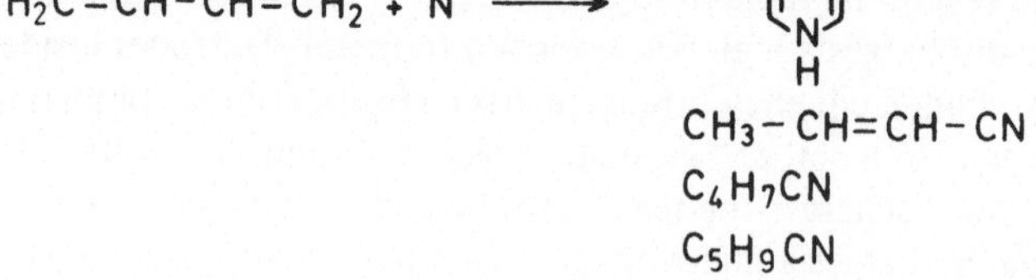

$$H_2C=CH-CH=CH_2 + N \longrightarrow$$

$$CH_3-CH=CH-CN$$
$$C_4H_7CN$$
$$C_5H_9CN$$

Atomic oxygen reacts with saturated organic compounds and abstracts hydrogen [6,10]. Unsaturated compounds add oxygen atoms to form epoxides, some of which rearrange to carbonyl compounds. From 2-pentene roughly equal amounts of epoxides, ketones and aldehydes are obtained [11]. The analogous synthesis of propylene oxide has been thoroughly studied because of its possible commercial interest [12]. The best results are obtained when oxygen and propene pass through a discharge together.

B. Isomerizations

Some 100–200 plasma reactions of organic molecules have been studied during recent years. This experimental material is not sufficient to establish general rules about the reactivity of organic molecules in plasmas. But it makes it possible to describe the plasma behaviour of several classes of compounds. Some typical reactions of organic compounds in glow discharges, like isomerizations, eliminations, dimerizations and polymerizations, have been studied in greater detail.

Numerous organic compounds isomerize when they pass through electrical discharges. Such rearrangements are frequently side reactions of little importance. In a number of cases, however, isomerization is the predominant reaction (Table 1). One example is the *cis-trans*-isomerization of olefines, a type of reaction which has been studied with stilbene [13]. Under mild reaction conditions *trans*-stilbene forms the *cis*-isomer as the sole reaction product. When the field strength is increased, a side reaction leading to phenanthrene gains in importance. Olefines which polymerize easily are difficult to isomerize and often give rather small yields.

Many isomerizations involve migrations of subsituents like the rearrangement of alkylaryl ethers to alkylphenols [14]. This reaction shows similarities to the Claisen rearrangement; however, primary, secondary, and tertiary alkyl groups migrate equally well. Anisol isomerizes to about 70 per cent in a single pass, forming *ortho*- and *para*-cresol. Besides these compounds, only phenol appears in significant amounts. In cresol methylethers the o-alkyl group migrates to the *ortho*- and *para*-positions [15].

Naphthyl methylethers are easily converted to methylnaphthols [15]. The α-isomer forms 2- and 4-methyl-α-naphthol, the β-isomer only 1-methyl-β-naphthol. Ethers with alkyl groups other than methyl sometimes show side reactions due to fragmentation of the migrating group. For example, *n*-propylphenylether yields mainly *ortho*- and *para-n*-propylphenol, but also some ethylphenols and cresols [14].

When diphenylether rearranges to hydroxybiphenyls, part of the ortho isomer cyclizes to dibenzofurane [16].

Similar rearrangements are observed with the nitrogen analogues [17]. N, N-dimethylaniline isomerizes to *ortho*- and *para*-N-methyltoluidines and is partly fragmented to N-methylaniline, benzonitrile and methyleneaniline. When N-methylaniline is the starting material, the main products are toluidines and aniline.

In addition to these rearrangements of more general importance, a number of isomerizations have been observed which apply only to specific cases. Some cycloolefines, like cyclooctatetraene and cycloheptatriene, are converted almost quantitatively to the alkylaromats styrene and toluene respectively [18]. Of possible interest for preparative chemistry are ring-chain isomerizations which have been observed with certain nitrogen compounds [19]. When pyrrole is subjected to a discharge it is largely converted to croton nitrile.

In a similar way indole isomerizes to benzyl cyanide, pyridine to cyano-butadiene and quinoline to cinnamic acid nitrile. Aniline and o-phenylenedi-amine also form large amounts of unsaturated nitriles in glow discharges.

$$\text{NH}_2 \text{-Phenyl} \longrightarrow \text{H}_3\text{C} - \text{CH} = \text{CH} - \text{CH} = \text{CH} - \text{CN}$$

Table 1. Examples of isomerizations in glow discharges [13-19]

Starting material	Conversion rate %	Products %
trans. Stilbene	20	*cis*-Stilbene 95
Anisol	67	Cresol, o:48; p:25
Phenetol	~30	Ethylphenol, o:41; p:29
N-Propyl-phenyl-ether	~30	Propylphenol, o:38; p:19
		Ethylphenol, o:1; p:0.5
		Cresol, o:7; p:4
1-Naphthyl-methyl-ether	13	Methyl-1-naphthol, 2:48; 4:35
2-Naphthyl-methyl-ether	88	1-Methyl-2-naphthol 45
Diphenyl-ether	40	Hydroxybiphenyl, 2:36; 4:18
		Dibenzofurane 9
N,N-Dimethylaniline	15	N-Methyl-toluidine, o:28; p:15
N-Methylaniline	6	Toluidine, o:28; p:6
Cyclooctatetraene	80	Styrene 40
Pyrrole	6	*cis-trans*-Croton nitrile 57
Quinoline	8	Cinnamic acid nitrile 33
		cis:5, *trans*:33

C. Eliminations

Plasma reactions which seem especially interesting for preparative work are eliminations. In many plasma reactions atoms or small groups are eliminated without destroying the rest of the molecule (Table 2). Thus aldehydes easily decarbonylate to the corresponding hydrocarbons [20]. The product obtained from benzaldehyde is mainly benzene and to a lesser extent biphenyl:

80%   20%

Pyridinealdehyde yields pyridine and some dipyridyl, a reaction which proceeds without destruction of the heterocyclic ring [19]. Ketones decarbonylate equally well. Benzophenone and dibenzoyl form diphenyl in high yields [20]. Several interesting eliminations have been found for cyclic and bicyclic ketones. In the case of cyclohexanone the decarbonylation accounts for only a small proportion of the reaction products. Benzoquinone decarbonylates to cyclopentadienone in good yields [20]. Under the reaction conditions this compound dimerizes and decarbonylates again.

The yields of decarbonylation reactions are quite high for bicyclic ketones. When camphor is slowly distilled through a glow discharge some 70–80 per cent of the starting material is converted predominantly to dimethyl-2,1,1-bicyclohexane [20]. Fluorenone decarbonylates to biphenylene almost quantitatively [21], thus providing a convenient one-step route for the preparation of

biphenylene which requires a sequence of several steps by standard chemical methods [22]. In a similar way, anthrone forms fluorenone with the loss of carbon monoxide.

Decarbonylations useful for preparative work are also observed with some aromatic hydroxy compounds. It is reasonable to assume a mechanism which involves the keto form [23]. Phenol, which shows little tendency to tautomerize, reacts mainly by forming benzene and to a small extent cyclopentadiene [14]. For

Table 2. Examples of elimination reactions [19-21,23]

Starting material	Conversion rate %	Products %
Benzaldehyde	63	Benzene 81
		Biphenyl 16
Pyridine-2-aldehyde	25	Pyridine 70
		2,2-Dipyridyl 20
Benzophenone	68	Biphenyl 70
Dibenzoyl	18	Biphenyl 98
Cyclohexanone	8	Cyclopentane 5
Camphor	29	Trimethyl-[2.1.1]-bicyclo-hexane 57
p-Benzoquinone	4	8,9-Dihydro-indenone (1) 60
Fluorenone	20	Biphenylene 98
9-Anthrone	69	Fluorene 77
1-Naphthol	20	Indene 76
2-Naphthol	24	Indene 92
Benzoic acid	20	Benzene 38
		Biphenyl 48
Phthalic anhydride	70	Biphenylene 60
		Triphenylene 27
Tetralene	50	Naphthalene 60
Tetrahydroquinolene	20	Quinolene 47
Acenaphthene	40	Acenaphthylene

naphthols the decarbonylation to indene becomes predominant while the elimination of the hydroxy group is of minor importance [23].

Another type of elimination reaction favoured under plasma conditions is the decarboxylation. Carbocyclic acids easily lose carbon dioxide to form the parent hydrocarbons. In acid anhydrides decarboxylation is followed by a decarbonylation. Cyclic or bicyclic anhydrides fragment forming unsaturated compounds, a reaction which has been studied with phthalic anhydride [24]. This anhydride decomposes to dehydrobenzene which, in the absence of other compounds, dimerizes, trimerizes or polymerizes. Orientation experiments indicated similar results for aliphatic acid anhydrides.

The elimination of hydrogen is very common in plasmas. Dehydrogenation of compounds like tetraline, acenaphthene, tetrahydroquinoline or indane has been carried out with excellent results. Elimination of hydrogen

Table 3. Examples of the formation of five-membered rings in glow discharges [17,20,25,26)]

Starting material	Conversion rate %	Products	Yield %
	~ 30		26
	68		27
	39		9
	40		30
	22		62
	24		82
	30		50
	13		57

halides or water has been studied only in a few examples. The dehydro-bromination of β-bromoethylbenzene and the dehydration of phenylethyl-alcohol both give high yields of styrene.

Interesting possibilities of synthesis arise when elimination leads to cyclic products. Such cyclization is fairly common with aromatic compounds. The mixture of products obtained from diphenylmethane contains about 25 per cent fluorene [25)] (Table 3). Similarly, benzophenone yields about 30 per cent fluorenone [20)]. The same method can be applied to the synthesis of hetero-cyclic compounds. Diphenylamine yields up to 30 per cent carbazol [26)], di-phenylether about 10 per cent dibenzofurane [16)].

In both cases rearrangement to substituted biphenyls followed by ring closure competes with direct cyclization. This assumption is supported by the fact that yields of carbazol and dibenzofurane are considerable higher when the starting material is 2-aminobiphenyl [26] or 2-hydroxybiphenyl.

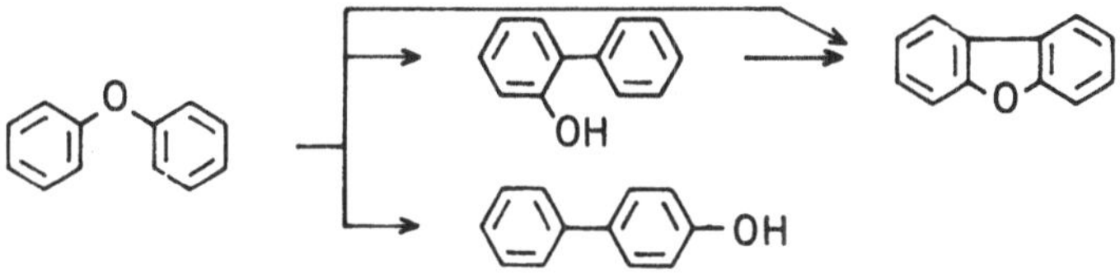

Nitrobiphenyl can also be converted to dibenzofurane. The nitrocompound, after isomerization to a nitrite, loses nitrogen oxide [27]. The remaining oxygen radical attacks the phenyl ring and cyclizes to the furane system.

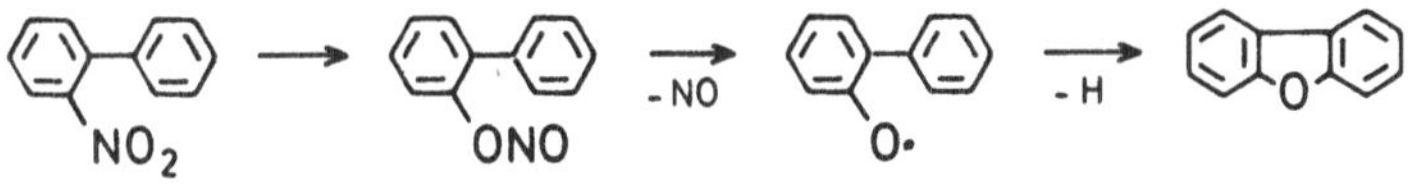

A number of examples are known where elimination gives six membered rings (Table 4). A carbocyclic ring is formed in the cyclization of stilbene to phenanthrene [13]. In a similar way stilbazol forms azaphenanthrene [19].

Six membered heterocyclics can be synthesized in various ways. Nitrogen analogues of stilbene, like the anilides of benzaldehyde or acetophenone sometimes form azaphenanthrenes [19]. Similarly the plasma reaction of azobenzene yields a small amount of diazaphenanthrene [19]. Better results are observed in the cyclization of hydroxy, amino or nitro compounds. Suitable starting materials are molecules with two phenyl groups linked by a carbon or hetero atom.

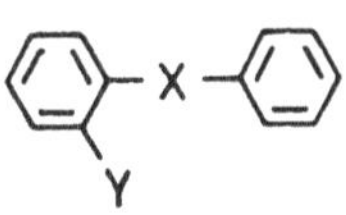

$X = CH_2 , CO , O , NH$

$Y = OH , NH_2 , NO_2$

49

Table 4. Examples of the formation of six-membered rings in glow discharges [13,16,18,19,25]

Starting material	Conversion rate %	Products	Yield %
	~ 30		19
	77		48
	29		3
	18		4
	80		3
	22		12
	90		45
	~25		~10
	93		14
	~30		~20

D. Bimolecular Reactions

In all the examples mentioned above the plasma process is a monomolecular reaction. Bimolecular processes can also occur but the experimental material presently available on such reactions is limited. A very simple type of bimolecular reaction involving only one compound is the dimerization of the starting material by loss of hydrogen or another group. Such reactions are frequently observed in plasmas. When benzene is subjected to a discharge the main product is biphenyl [28-31]. Sometimes the reaction proceeds further to terphenyl.

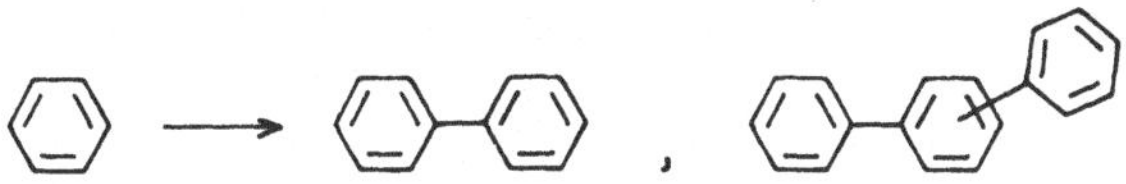

The dimerization of toluene and substituted toluenes leads to diaryl-ethanes [25,28,32]. Electron-attracting substituents favour the reaction [32] while electron-donating groups reduce the yield. In alkylaromats with straight or branched alkyl groups it is almost always the weakest bond of the side chain which is broken [25,28]. The resulting radicals combine to diarylethanes or substituted diarylethanes:

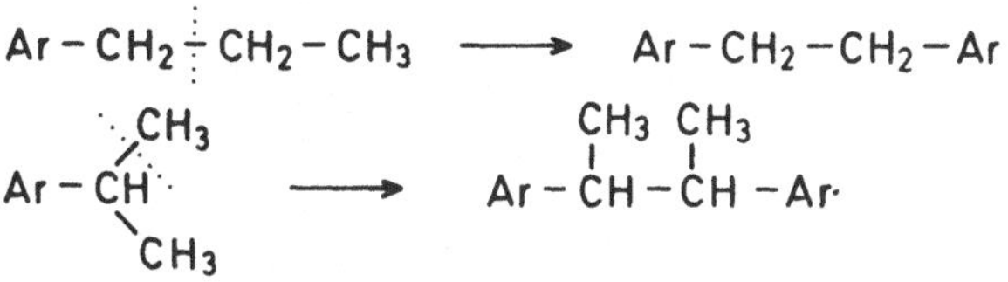

Aliphatic compounds dimerize less readily. Methane forms considerable amounts of ethane but also various other saturated and unsaturated hydrocarbons [33,34,40]. The dimerization of propene and substituted propenes gives mainly hexadienes.

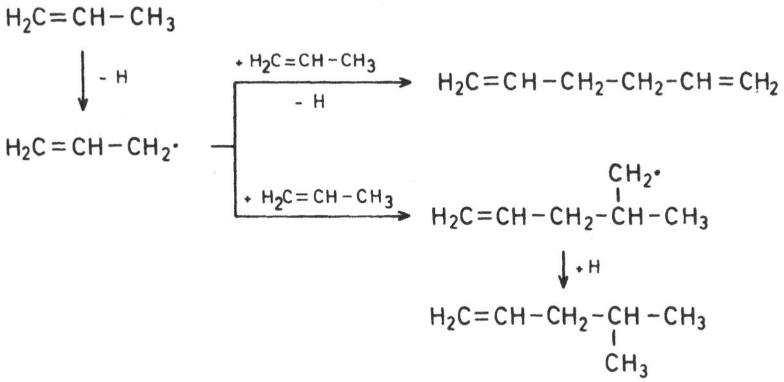

However the propenyl radicals also add to the double bond of other propene molecules. From isobutylene about equal amounts of dimethylhexadiene and dimethylhexenes are obtained.

51

Plasma polymerizations are also very common [30,35-38]. It is possible to convert almost every organic compound into a polymer by using vigorous reaction conditions [35,39]. Even under mild conditions various olefines or dienes polymerize readily, either in the gaseous phase or on the surrounding walls. This makes it possible to coat various objects with thin polymer films, a technique which can be used to modify surfaces.

Plasma polymerizations allow some interesting variations when the monomers are themselves products of plasma reactions, for example, the fragmentation of phthalic anhydride to dehydrobenzene which under certain reaction conditions can polymerize to polyphenyl [24]. Also, the isomerization of some nitrogen heterocyclics gives unsaturated nitriles which polymerize easily [19].

Very few cases of oligomerization have been observed in discharges. A rather surprising example is offered by the experiments with acetylene [18]. This compound has been studied frequently with various kinds of discharge, the product being always a yellowbrown polymer (cuprene) [39,41,42]. However, when the surface area of the reaction vessel is large, polymerization can be prevented and compounds having four, eight or ten carbon atoms are isolated, predominantly styrene, phenylacetylene, naphthalene and benzene [18].

Plasma reactions with two or more compounds require careful measurement of all substances in order to ensure the correct total pressure and the required ratio of reagents. Since such reactions are difficult to optimize, little experimental material is at present available.

In some experiments one constituent acts mainly as a dilution agent, *i.e.* when an inert gas is used as carrier for the organic material. The inert gas reduces the number of collisions between organic molecules and in many cases inhibits polymerization. In other cases the inert gas molecules, having low-lying excited states, transfer energy to the organic compounds by collision.

Reactions between two different molecules have been studied in great detail for atomic oxygen and propene [12]. Mixtures of oxygen and benzene form some phenol in glow discharges [23,43]. Similarly, mixtures of benzene and ammonia yield aniline [19]. Since the latter reactions have not been optimized, it is not known whether they can compete with standard chemical methods.

In other types of bimolecular reactions reactive intermediates are trapped with various additives. These synthesic possibilities have been studied with phthalic anhydride [24]. The compound decomposes in two steps to dehydrobenzene which in turn dimerizes in two steps via a biphenyl radical:

Various additives can scavenge these intermediates. Added hydrogen traps the diphenyl diradical, while acetylene combines with the dehydrobenzene. Ammonia forms addition compounds with all three intermediates to give aminobenzaldehyde, benzamide, aniline and carbazol.

III. Reaction Mechanisms

At present little is known about the mechanism of plasma reactions, due to the complexity of the system [44-45]. Electrons, positive and negative ions, excited species and radicals are all present in plasma in addition to the molecules of the starting materials and the products.

Most plasma reactions involve excited molecules which are generated through direct collision, via negative ions, or by the recombination of positive ions with electrons. The excited molecules can fragment or isomerize either to stable compounds or to reactive intermediates which are precursors of the final products. Frequently, excited molecules, ions, or intermediates attack surrounding molecules so forming the products in bimolecular processes. An estimate of the importance of the various elementary processes would require the determination of the energy distribution and concentration of all charged and neutral species. Such measurements are complicated and can be carried out only in rare cases.

Some general ideas about the reaction mechanisms involved can be obtained from a comparison of plasma products with the products of other reactions. Often similarities are found between plasma chemistry and photochemistry, pyrolysis, radiochemistry or mass spectroscopy. If the reaction products are identical, it is reasonable to assume that the reaction mechanism will be the same or at least the final steps in a sequence of elementary processes. Occasionally, the products indicate mechanisms which have no known analogues in other fields of organic chemistry [46].

Few mechanisms of plasma reactions have been studied in any detail. In the reaction of toluene the bimolecular dimerization to bibenzyl competes with the monomolecular generation of methyl and phenyl radicals which are

the precursors of benzene and ethylbenzene [25]. The rearrangement of alkylaryl ethers to alkylphenols was found to proceed by a chain mechanism
having a phenoxonium ion as the chain propagator [17]. The migration of the
alkyl group occurs within an ion-molecule complex. The oligomerization of
acetylene probably starts with the dimerization of acetylene to cyclobuta-
diene [18]. This intermediate forms vinylacetylene and diacetylene by ring
opening and with acetylene, vinylacetylene and diacetylene forms the addition compounds benzene, styrene and phenylacetylene.

In plasma experiments only the starting materials and final products are
known. It is rare for intermediates to be identified with certainty so that
many questions concerning the mechanism remain open.

IV. Outlook

The recent investigations of organic compounds in glow and corona discharges
have indicated numerous possibilities for preparative applications and furthermore provided interesting information concerning reaction mechanisms. Most
studies have been carried out with simple systems. Aromatic compounds
proved to be particularly suitable models because of the great stability of the
aromatic ring. The results obtained with these compounds generally apply
to nonaromatic substances, too.

Not every compound is suitable for plasma experiments. Since plasma is
limited to the gas phase, only compounds which distill in vacuum without de-
composing can be used in plasma chemistry. When such molecules have several
reactive groups, they frequently yield a number of products. If, however, one
of the possible reaction routes requires considerably less energy than the others,
the reaction leads exclusively or predominantly to a single product.

Plasma chemistry is at present mainly an empirical technique. Little is
known about relationships between the properties of the plasma and the reaction mechanism or the product distribution. Experimental conditions have to
be optimized for every new reaction. Once more experimental material and
improved methods of diagnosing plasma become available it may become routine to adjust plasmas for each specific chemical problem.

V. Literature References

[1] Bondt, Deiman, Paats von Troostwyk, Lanwerenburg: Ann. Chim. *21*, 48 (1796).
[2] Berthelot, M: Compt. Rend. *67*, 1141 (1869). − *82*, 1283 (1876).
[3] McTaggart, F. K.: Plasmachemistry in electrical discharges, Chap. 10. Amsterdam:
Elsevier 1967.
[4] Lemmon, R. M.: Chem. Rev. *70*, 95 (1970).
[5] Miller, S. L.: J. Am. Chem. Soc. *77*, 235 (1955). − Science *117*, 528 (1953).

[6] Wagner, H. Gg., Wolfrum, J.: Angew. Chem. *83*, 561 (1971).
[7] Wright, A. N., Winkler, C. A.: Active Nitrogen. New York: Academic Press 1968.
[8] Tsukamoto, A., Lichtin, N. N.: J. Am. Chem. Soc. *82*, 3798 (1960). − J. Am. Chem. Soc. *84*, 1601 (1962).
[9] Fujino, A., Lundsted, S., Lichtin, N. N.: J. Am. Chem. Soc. *88*, 775 (1966).
[10] Blaustein, B. D., Fu, Y. C., in: Organic reactions in electrical discharges. Techniques of chemistry in physical methods of chemistry (ed. A. Weissberger, B. W. Rositer), Part II. New York: Wiley-Interscience 1971.
[11] Cvetanovic, R. J.: J. Chem. Phys. *25*, 376 (1956). − J. Chem. Phys. *30*, 19 (1959). − Can. J. Chem. *36*, 623 (1958).
[12] Weisbeck, R., Hüllstrung, R.: Chem. Ingr.-Tech. *42*, 1302 (1970).
[13] Suhr, H., Schücker, U.: Synthesis *1970*, 431.
[14] Suhr, H., Weiß, R. I.: Z. Naturforsch. *25b*, 41 (1971).
[15] Suhr, H., Weiß, R. I.: Liebigs Ann. Chem., in press.
[16] Weiß, R. I., Suhr, H.: Publication in preparation.
[17] Weiß, R. I., Suhr, H.: In press.
[18] Suhr, H., Rosskamp, G.: Publication in preparation.
[19] Suhr, H., Schöch, U.: Publication in preparation.
[20] Suhr, H., Kruppa, G.: Liebigs Ann. Chem. *744*, 1 (1971).
[21] Suhr, H., Weiß, R. I.: Angew. Chem. *82*, 295 (1970). − Int. Ed. *9*, 312 (1970).
[22] Lothrop, W. C.: J. Am. Chem. Soc. *63*, 1187 (1941); − *64*, 1698 (1942).
[23] Suhr, H., Szabo, A.: Publication in preparation.
[24] Suhr, H., Szabo, A.: Liebigs Ann. Chem. *752*, 37 (1971).
[25] Suhr, H.: Z. Naturforsch. *23b*, 1559 (1968).
[26] Suhr, H., Schöch, U., Rosskamp, G.: Chem. Ber. *104*, 674 (1971).
[27] Suhr, H., Rosskamp, G.: Liebigs Ann. Chem. *742*, 43 (1970).
[28] Suhr, H., Rolle, G., Schrader, B.: Naturwissenschaften *55*, 168 (1968).
[29] Stille, J. K., Sung, R. L., van der Kooi, J.: J. Org. Chem. *30*, 3116 (1965).
[30] Ranney, M. W., O'Connor, William F.: Chemical reactions in electrical discharges. Advances in chemistry. Series 80, p. 297. Washington DC: Am. Chem. Soc. 1969.
[31] Weisbeck, R.: Chem. Ingr.-Technik *43*, 721 (1971).
[32] Suhr, H., Schöch, U., Schücker, U.: Synthesis *1971*, 426.
[33] Il'in, D. T., Eremin, E. N.: Organic reactions in electrical discharges (ed. N. S. Pechuro), pp. 11, 22, 25. Russ. original Moskau: Nauka Press 1966, Engl. transl. New York: Consultants Bureau 1968.
[34] Borisova, E. N., Eremin, E. N.: Organic reactions in electrical discharges (ed. N. S. Pechuro), p. 52. Russ. original Moskau: Nauka Press 1966, Engl. transl. New York: Consultants Bureau 1968.
[35] Wightman, J. P., Johnston, N. J.: Chemical reactions in electrical discharges. Advances in chemistry. Series 80, p. 322. Washington DC: Am. Chem. Soc. 1969.
[36] Neiswender, D. D.: Chemical reactions in electrical discharges. Advances in chemistry. Series 80, p. 338, Washington DC: Am. Chem. Soc. 1969.
[37] Secrist, D. R.: Chemical reactions in electrical discharges. Advances in chemistry. Series 80, p. 242. Washington DC: Am. Chem. Soc. 1969.
[38] Hollahan, J. R., McKeerer, R. P.: Chemical reactions in electrical discharges. Advances in chemistry. Series 80, p. 272. Washington DC: Am. Chem. Soc. 1969.
[39] McTaggart, F. K.: Plasma chemistry in electrical discharges, p. 196. Amsterdam: Elsevier 1967.
[40] Blaustein, B. D., Fu, Y. C., in: Organic reactions in electrical discharges. Techniques in physical methods of chemistry (eds. A. Weissberger, B. W. Rossiter), Part II, p. 153. New York: Wiley-Interscience 1971.

[41] Il'in, D. T., Eremin, E. N.: Organic reactions in electrical discharges (ed. N. S. Pechuro), p. 8. Russ. original Moskau: Nauka Press 1966, Engl. transl. New York: Consultants Bureau 1968.
[42] Borisova, E. N., Eremin, E. N.: Organic reactions in electrical discharges (ed. N. S. Pechuro), p. 33. Russ. original Moskau: Nauka Press 1966, Engl. transl. New York: Consultants Bureau 1968.
[43] Chu, J. C., Ai, H. C., Othmer, D. F.: Ind. Eng. Chem. *45*, 1266 (1953).
[44] Tal'rose, V. L., Karachertsev, G. V.: Reactions under plasma conditions (ed. M. Venugopalam), Vol. II, p. 35. New York: Wiley-Interscience.
[45] Polak, L.: Reactions under plasma conditions (ed. M. Venugopalam), Vol. II, p. 141. New York: Wiley-Interscience.
[46] Suhr, H., Rosskamp, G.: Liebigs Ann. Chem. *742*, 43 (1970).

Received May 4, 1972

The Infra-Red Spectra of Crystalline Solids

Prof. C. J. H. Schutte

Department of Chemistry, University of South Africa, Pretoria, South Africa

Contents

1. Introduction

Preliminary

The advent of the modern generation of computers has made it possible to calculate the frequencies of vibration and rotation of simple isolated inorganic and organic molecules using the so-called *ab-initio* methods of quantum mechanics [1,2]. This means that quantum mechanics has confirmed the fact that the infra-red and Raman spectra of isolated molecules are due to the excitation of isolated molecules between quantized vibrational and rotational energy levels by electromagnetic radiation. The calculation methods have been perfected in such a way that actual observed spectra can be matched against computed spectra, thus identifying the absorbing species absolutely. A case in point here is, *e.g.* the H_2-calculation by Wolniewicz [3], as well as the predicted spectra of LiCN and LiNC by Bak, Clementi and Kortzeborn [4].

This concept of isolated molecules undergoing vibrations and rotations when excited by electromagnetic radiation of the appropriate Bohr frequency, is applicable only to rarified pure gases. As soon as the pressure rizes, collision becomes very important and the electronic structure of a molecule perturbs the electronic structure of the other member of a colliding pair. It follows that such a perturbed molecule will have an "equilibrium geometry" which is different from that of the unperturbed molecule, and hence the potential energy hypersurface opposing vibrations changes. This means that the resulting rotation-vibration spectrum is different from that of the unperturbed molecule. The change is usually small, so that as the pressure rizes the rotation-vibration band becomes smeared-out and the rotational fine structure is lost. This type of effect is clearly illustrated by the spectrum of HCl (g) compressed by N_2 (g) where the prohibited Q-branch even appears as obtained by Coulon *et al.* [5a]. As the collision frequency increases, there is always the chance that a collision pair exists long enough so that they can act as an "activated molecule". Such an effect is observed in the simultaneous transitions observed in liquids first found in liquid mixtures by Ketelaar and Hooge [5c], *e.g.* in mixtures of $CS_1 (1)$ and $Br_2 (1)$ the transitions $\nu_{CS_2} \pm \nu_{Br_2}$ occur where the "activated molecule" absorbs one quantum of radiation (these simultaneous are also observed in gases [5b]). The structure of such an "activated molecule" is still very near to that of the two unperturbed molecules because the frequencies $\nu_{CS_2} \pm \nu_{Br_2}$ are very near to the sums of the pure compounds (also liquids!), but stable molecules can also be formed, *e.g.* by hydrogen bonding in CH_3COOH which form dimers (see *e.g.* Witkowski [6]) with completely different vibrational spectra.

In crystals the colliding pairs of gases and liquids are replaced by identical molecules in a regular repetitive *close-packed* array which forms a lattice consisting of unit cells (see, for example, any text-book on crystallography). This means that there can be no (or very little) rotational freedom for the molecules, and that the fleeting collision perturbation in gases now becomes, as it were, "frozen in" and the whole ensemble of molecules must be considered to be one vast molecule stretching over the volume of the crystal. Each molecule is distorted from its "free" geometry by the average perturbation of the molecules surrounding it (both short and long range forces operate here), and each molecule is coupled to its neighbours through this surrounding averaged perturbing field. If the surrounding field is small, then each distorted molecule absorbs as if it alone were present in the lattice; this situation is described by the *site-group method* of Halford [7] (see also Couture [8] for corrections). If the surrounding perturbing field is large, it means that all the molecules are coupled together (like coupled oscillators); the *unit-cell method (factor-group method)* of Hornig [9] adequately describes this coupling and treats the atoms contained within one unit cell as if they constitute a unit-cell molecule. Another feature, peculiar to the solid state, is the occurrence of lattice modes of long wave-length due to the translations of molecules with respect to each other, and to "rotations" which are caused by vibratory twisting motions in the lattice; both types of motion are called *external modes.*

Both the unit-cell group method and the site-group method have recently been discussed by Carter [10] and especially clearly (with many worked-out examples) by Fateley, McDevitt and Bently [11], so that it is unnecessary to re-

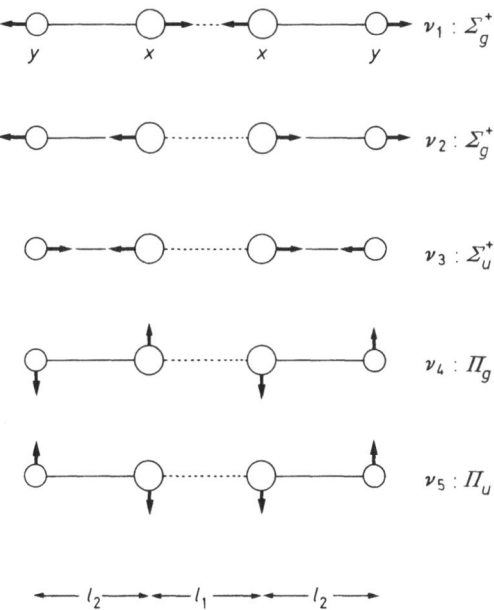

Fig. 1. The coupling case Y-X$\cdots$X-Y as discussed in text

peat a detailed description here. Bertie and Bell [12] have given a critical discussion of problems relating to these methods.

Both methods can easily be derived from a very simple model. Consider a
unit cell which contains two XY molecules which are equivalent in the group-
theoretical sense, *i.e.* they are transformed into one-another by the operations
of the group of the *unit cell* (this group is the *factor group* of the space group
[10] and is *isomorphous* with one of the 32 crystallographic point groups).[a]
An idealized situation is shown in Fig. 1. If k_1 is the force constant of the XY
bond, k_2 that of the X$\cdots$X bond, and k_δ the force constant corresponding
to the change in angle between XX and YY, then the frequencies of the normal
modes which are very schematically given (following Herzberg [13]) in Fig. 1,
are as follows

$$\nu_1 + \nu_2 = 2\frac{k_1}{m_x} + \mu k_2; \quad \nu_1\nu_2 = 2\frac{k_1 k_2}{m_x m_y},$$

$$\nu_3 = \mu k_2,$$

$$\nu_4 = \frac{k_\delta}{l_1^2 l_2^2}\left[\frac{l_1^2}{m_y} + \frac{(l_z + 2l_2)^2}{m_x}\right],$$

$$\nu_5 = \mu\frac{k_\delta}{l_2^2}; \quad \mu = \left(\frac{m_x + m_y}{m_x m_y}\right) \quad i.e. \text{ inverse reduced } xy \text{ mass}$$

assuming valuence forces (see Herzberg[13],p. 181).If k_1 is assumed to be independent of k_2, then ν_i are smooth functions of k_2 and k_δ. If $k_2 \to 0$ (*i.e.* no
interaction) then $\nu_3 \to 0$, $\nu_1 + \nu_2 \to 2k_1/m_x$ and $\nu_1\nu_2 \to 0$; making the logical
choice that $\nu_2 \to 0$ (because there is no internal potential opposing the motion),
then $\nu_1 \to \nu_{xy}$ of the "free" XY group imbedded in the general field of the
crystal (which perturbs the frequency from the ideally free value). If both
$k_2, k_\delta \to 0$, then $\nu_4, \nu_5 \to 0$, indicating that there is no potential opposing the
rotational motion of the "free" XY groups. In this case ν_2 becomes one of the
translational (external) motions and ν_5, ν_6 become part of the rotational motions (external modes) of the crystal; ν_3 is the antisymmetric stretching mode,
and ceases to have any meaning in the absence of any coupling (destructuve
interference within the same unit cell). It also follows, when $k_2 = 0$, that the
symmetric stretching mode $\nu_1 > \nu_{xy}$ (eliminating ν_2 from the first two equations of Eq. (1) and taking the positive square root), $\nu_2 < \nu_1$ if $k_2 < k_1$ and
$\nu_3 < \nu_1$. This simple model thus predicts two stretching modes for XY in the

---

[a] See, however, Bertie and Bell [12] about the definition of the terms factor group and
unit-cell group, and references therein.

simple lattice chosen; the difference $\nu_1 - \nu_3 \geqslant 0$, depending upon k_2. The site-group approach of Halford [7] assumes k_2 and k_δ equal to zero, so that the site group in this case becomes $C_{\infty v}$; the frequency of the XY molecule as perturbed ($H = H^0 + H'$) by the general crystalline field, is ν_{xy} which is different from ν_{xy}^0 of the totally free ion where $H' \equiv 0$. The factor-group approach uses the "factor group" $D_{\infty h}$ and if k_2, $k_\delta = 0$ predicts two stretching frequencies viz. ν_1 (symmetric stretch) $\sim \nu_{xy}$ and ν_3 (antisymmetric stretch) $< \nu_{xy}$, ν_1.

One of the most important points, adequately emphasized by Fateley, McDevitt and Bently [11] in their excellent paper, is that the *spectroscopic unit cell* must be chosen, and *not* the crystallographic unit cell. The spectroscopic unit cell is identical with the Bravais cell; the Bravais cell is identical with the crystallographic unit cell in the P (primitive) cells, but the other structures, designated by B, C, I, F . . . . . in the crystallographic tables [14] contain 2, 2, 3, 4, ... Bravais unit cells. Fateley, McDevitt and Bentley [11] describe the procedure to obtain the correct Bravais (spectroscopic) unit cell in detail, and should be consulted, and it must be remembered that a R-unit cell is either three times larger than the spectroscopic unit cell, or identical to it, depending upon the specific case. Adams and Newton [15,16] published a paper which contain tables for the factor group analyses of the vibrational spectra of solids.

2. The Scope of this Review

The use of the infra-red spectrometer as a routine instrument has resulted in an explosion of published infra-red spectra. As it is impossible to list and discuss the infra-red spectra of even such a "restricted" field as the solid state, it is inevitable that a somewhat arbitrary choice must be made in order to comply with the size of this review. The main emphasis is upon the published work during the five-year period 1966–1970, although some important earlier and later papers will also be dealt with. Only papers which give a full vibrational analysis of at least the internal modes of the constituent ions/molecules in the crystal lattice will be included, and full use will be made of the evidence of both infra-red and Raman spectra. This means that papers which "just report" the spectrum of a given compound, or give only a partial analysis for "finger-printing" purposes, will not be included, unless they are especially significant; this excludes the great majority of the papers on inorganic and organic solids.

The following will be the main sections of solid-state infra-red spectroscopy with which this article will concern itself:

the application of the factor group method to: (i) simple inorganic compounds and (ii) to simple organic compounds;

condensed pure gases; solid solution and matrix spectroscopy;
complexes; infra-red dichroïsm; the study of phase transitions;
internal reflection spectroscopy; pressure effects in spectra.

It is impossible to discuss all papers which have appeared on the subjects
included in this review (in order to keep it within reasonable limits); most of
the important papers are included, and this review should give the reader an
insight in the state of the art of solid state infra-red spectroscopy. The attention of the reader is drawn to the existence of *Science Citation Index*[a]; if a
reference of this paper is used as a leading reference, *Science Citation Index*
will quickly yield a list of other works which have cited the particular reference, enabling him to do a relatively complete literature search in a short
time.

Previous reviews on solid-state infra-red spectroscopy are, *e.g.* by Vedder
and Hornig [17] which contains a relatively complete bibliography of all the
most important work up to 1960 and by Mitra [18], while Wilks and Hirshfeld [19] have written a review of internal reflection spectroscopy. Orville-
Thomas [20] wrote a review on I. R. spectroscopy as a diagnostic tool, and
Annual Reviews of Physical Chemistry [21] often contain review papers on
infra-red spectroscopy which also refer to the solid state.

3. The Application of the Factor Group Method

3.1. Applications to Small Inorganic Compounds

Very many such compounds containing relatively few equivalent (in the group-
theoretical sense) molecules or ions in the spectroscopic unit cell have been
exhaustively investigated by infra-red spectroscopy and the resulting spectra
interpreted by means of the correlation between the point group of the free
molecule and the factor group of the unit cell *via* the site group. However,
many authors did not realize the significance of the difference between the
crystallographic and spectroscopic unit cell groups, and included too many
vibrations to describe the vibrations of the lattice. In what follows, reference
will only be made to the internal motions of molecules or ions or lattices, except for a few cases where libration or "free" rotation occurs.

a) Planar XY₃ Ions in Lattices

The planar XY_3-ions like NO_3^- (nitrate) and CO_3^- (carbonate) are perennial
favourites because the D_{3h}-structure is especially sensitive to the perturbing
crystalline field. The following normal modes are generated by the D_{3h}-struc-

---

[a] "Science Citation Index", Institute for Scientific Information, Inc., Philadelphia.

ture: A_3' (R), A_2' (IR), $2E'$ (R + IR), as discussed by, *e.g.* Buijs and Schutte [22] and Schroeder, Weir and Lippincott [23]; any site symmetry lower than that of the subgroup C_{2v} of D_{3h} will automatically split the E-modes of the XO_3 ion so that the ion is a very sensitive probe for lattice forces or even crystal-surface effects.

The surface of an ionic crystal is structurally different from that of the bulk solid. Weil [24] contends that the surface extends for *ca.* 10,000 atom layers; this means that the surface of a crystal, as examined by attenuated reflection spectroscopy, should show signs of this difference. Cooney *et al.* [25] took this idea up, and used the ν_2-mode of NO_3^- in $LiNO_3$ and $NaNO_3$ to detect this effect; the "surface depth" found is far greater than that predicted theoretically, and is dependent upon temperature (up to the melting point). Molten nitrates are studied very much in order to determine what the nature of the liquid state is, *e.g.* Devlin, James and Frech [26] and references therein, James and Leony [27], and Brooker [28].

The nitrates of the alkali metals, the alkaline-earth metals and $AgNO_3$, $TlNO_3$, etc. have been extensively studied again, and it is regrettable that so many papers contain incomplete lists of references. Brooker [28] reported new spectral features in the out-of-plane deformation and its first overtone of a large number of nitrates, using Raman spectra of polycrystalline samples, and band multiplicity in the overtone region has been used to confirm the non-coincidence between the Raman and I.R.-components of the fundamental; this paper gives a useful summary of the crystal structures of several nitrates. The isomorphic cubic nitrates of Pb, Sr, Ba have been extensively re-examined; Schutte [29] report that the unit-cell group method indicates that the newly-reported Birnstock non-centro symmetric space group [30] T^4-$P2_1^{x3}$ with $Z = 4$ could be correct, but Brooker, Irish and Boyd [31], and Schutte in an earlier paper [32] show that an adequate explanation is afforded by the factor-group splittings of the centrosymmetric Vegard space group T_h^6-$P2_1^x$ $3i$ with $Z = 4$. There is, however, still some controversy regarding the crystal structure of $Ba(NO_3)_2$ as shown by the work of Sirdeshmulch [33] and Belyustin, Levina and Novoset'tseva [34], so that the problem must be regarded as still open to verification. Bon and Vergnoux [35] studied these nitrates at various temperatures using single crystals.

The Raman and infra-red spectra of *single crystals* of the alkali-metal nitrates and $AgNO_3$ have been carefully studied by James and Leony [36], who give tables of factor group splittings predicted for all the major nitrate crystal structures. This study is amplified by the work of Miller *et al.* [37] on $LiNO_3$ (polarized Raman Study; his ν_1 of 1071.0 cm^{-1} is in agreement with James and Leony [36], but appreciably smaller than previous values of *ca.* 1086 cm^{-1}); the absence of ν_3 and ν_4 from the $x(zy)y$ spectrum is interpreted as evidence for the absence interionic coupling. Nakagawa and Walter [38] have executed a normal coordinate analysis for NO_3^- in $LiNO_3$, $NaNO_3$ and KNO_3, obtaining reasonable agreements between calculated and observed values. Rosseau, Miller and Le Roi [39] reported the single crystal Raman spectrum of $NaNO_3$ which is birefringent (the problems of birefringent crystals have been discussed by Porto, Giordmaine and Damen [40] in their Raman Study of $CaCO_3$ calcite).

The power of the reflection method is illustrated by the important study of Hellwege *et al.* [41] on the two-phonon absorption spectra and dispersion of the phonon branches in crystals of the calcite type, that is $NaNO_3$, $CaCO_3$, $MgCO_3$ (with space group C_{3d}^2-$R\bar{3}c$) and dolomite $CaMg(CO_3)_2$ (with space group C_{3i}^2-$R\bar{3}$) and dolomite $CaMg(CO_3)_2$ (with space group C_{3i}^2-$R\bar{3}$). The spectra were recorded from $4000-20$ cm^{-1} using both absorption and polarized reflection spectra (for the theory of reflection spectra, see Vedder and Hornig [17]); extensive tables of reflection and absorption spectra are given.

Most papers dealing with the spectrum of the carbonate ion CO_3^{2-} neglect to mention the important paper by Decius, Malan and Thompson [42] on the effect of intermolecular forces on molecules in the crystalline state which refer specifically to the out-of-plane bending mode of CO_3^{2-}. In this paper they derive the dependence of this mode upon the ^{12}C-^{13}C isotopic ratio. Sterzel and Chlorinski [43] also discuss the effect of isotopes upon the CO_3^{2-} vibrations; these two papers should be consulted when assigning CO_3^{2-}-spectra because the modes depend very much upon the ^{12}C-^{13}C ratio. Orville-Thomas [20] has discussed the dependence of the CO_3^{2-} force constants upon the C–O distance, and shows that this leads to a bond intermediate between a single and a double bond.

The CO_3^{2-}, CS_3^{2-} and CSe_3^{2-} force constants are discussed by Müller and Fadini [44]; this is also discussed for NO_3^- by Becher and Ballein [45] in a paper which eliminates the multiplicity in the calculation of force constants from normal vibrations, and Pfeiffer [46] gave a simpler method to calculate this.

The formate ion HCO_2^- is also planar, and exhibits a relatively simple vibrational structure [47-52] in simple alkali-metal formates. Kuroda and Kubo [53] recently studied the normal vibrations of copper(II)formate tetrahydrate, copper(II)formate tetradeuterium oxide, as well as anhydrous $Cu(HCO_2)_2$ from $4000-200$ cm^{-1}. The tetrahydrate has a layer structure, which was analysed with D_{4h}^5 *via* the factor-group method using the method of Bhagavantam and Venkaterayudu [54].

b) The Hydroxyl Group

The isolated hydroxyl group shows only one infra-red absorption mode due to the O–H stretching but compounds containing ionic hydroxyl show a surprising amount of detail; a case in point here is mica as discussed by Mitra [18], and reviewed by Buchanan, Caspers and Murphy [55]. The same effect is also shown by the spectrum of $Ca(OH)_2$ and $Ca(OD)_2$ as reported by Oehler and Günthard [56] for room temperature and for liquid He-temperature. Oehler and Günthard use a method they developed [57] to determine the symmetry coordinates and factorization of vibrational problems for crystals with symmorphic space groups to describe the OH-vibrations in $Ca(OH)_2$; in paticular the OH-stretching region is discussed in terms of transitions between critical points of the dispersion surface following a suggestion of Mitra [58]. Oehler and Günthard [59] followed this up with a paper on the symmetry property of the critical point which is responsible for the combination bands in the OH stretching region of $Ca(OH)_2$.

Following their paper on the microwave spectrum of CsOH(g) [60], Acquista, Abramonitz and Lide [61] studied the spectra of matrix-isolated (in argon) CsOH and CsOD; the Cs–O stretching mode is 335.6 and 330.5 cm^{-1} and the bending mode at 306 and 226 cm^{-1}, respectively for CsOH and CsOD, indicating a linear, or nearly-linear structure (which is also found in RbOH, RbOD, NaOH, NaOD [62]. The OH radical was trapped by Acquista, Schoen and Lide [63] in rare-gas matrices at ~4 °K by flash photolysis and they report an OH-stretching of ~450 cm^{-1}, but did not observe any rotational behaviour, although two different frequencies were observed. The libration of the OH⁻ ion substitutionally imbedded in alkali-halide crystals was studied by Klein, Wedding and Levine [64]; they report a librational mode of ~300 cm^{-1} in the various alkali-halides (the exact frequency depending upon the particular halide), which also occurs in the form of a combination band *ca.* 300 cm^{-1} above the OH-stretching mode at ~3600 cm^{-1}. The Devonshire model [65] which describes the quantum mechanical rotation of a diatomic molecule in an octahedral hole, is used to explain the OH-OD frequency differences. The same problem of OH⁻ ions in KCl and NaCl lattices was attacked by Scott and Flygare [66] using microwave spectroscopy; they detected several lines between 8–40 GHz. The zero-field spectra obtained indicate that the OH⁻ ion experiences a strong C_{4v} perturbing potential in addition to the expected octahedral field; the C_{4v} field probably arizes from the shifting of the OH⁻ mass centre from the lattice site in question. The OH⁻ is discussed in terms of a tunneling librator (see Refs. [67,68]) instead of a hindered rotor, as done by Devonshire [65]; a correlation diagram between a free rotator and a tunneling vibrator is given.

The infra-red spectrum of Li-6 hydroxide was discussed by Decius and Lilley [69] who report that the major feature at 3681 ± ~400 cm^{-1} in the LiOH spectrum (which was previously assumed to involve a librational mode) is partly due to the absorption at 3681 ± ~290 cm^{-1} which involves a translatory mode in which there is a significant amount of Li^7-participation; the rest may be a two-phonon process (see Oehler and Günthard [56]).

c) Tetrahedral XO_4-Species

Very many papers referring to mineral silicates, etc. have appeared, but they will not be referred to here, except for the illuminating paper of Oehler and

Günthard [70] on the low-temperature infra-red spectra of silicates with the olivine Mg_2SiO_4 structure (see also Tarte [71,72]) which also includes a normal-coordinate analyses. Tarte [73], also published an excellent paper on the structures of the aluminates containing the AlO_4-tetrahedra (see also Carrerra *et al.* [74] for IR and Raman spectra of $Al(OH)_4^-$ ions), and Griffith [75] analysed silicate spectra.

Compounds containing PO_4^{3-} groups and the various hydrophosphates, *e.g.* HPO_4^{3-} etc. continue to be extensively investigated. Ratajczak [76] investigated the ferroelectrics KH_2PO_4 and $NH_4H_2PO_4$ by polarized infra-red radiation; Coignac [77,78] also investigated KH_2PO_4, as well as Hill and Ichiki [79]; these investigations give some insight into the nature of the ferroelectric phases.

Kravitz, Kingsley and Elken [80] and Kingsley, Mahan and Kravitz [81] made a Raman and IR study of hexagonal $Ca_{10}(PO_4)_6F_2$ and discussed the relevant splittings and Davydov-splittings (or factor group) [82,83].

The sulphate ion SO_4^{2-} also continue to be studied and Ananthanarayanan [84] published a paper on the SO_4^{2-}-symmetry in crystals, referring especially to single crystals of the double sulphates $M_2'M''(SO_4)_2 . 6H_2O$ where $M' = K^+$, NH_4^+ and $M'' = Mg^{2+}$, Zn^{2+}, Ni^{2+} or Co^{2+}. Hester and Krishnan [85] discussed the IR spectra of molten sulphates.

The permanganate ion MnO_4^{2-} is also drawing some attention. Hendra [86] studied the Raman spectrum of $KMnO_4$, $AgMnO_4$ and $CsMnO_4$, and suggested that MnO_4^- is always contaminated by MnO_2. Doyle and Kirkpatrick [87] investigated the IR spectra of nine hydrated and ten anhydrous permanganates, discussing the results in terms of the site symmetry of the MnO_4^- group. It might be worth-while to study these permanganates at low temperatures, to "sharpen-up" some of the details, so that the factor-group correlations can be used, since most crystal structures are sufficiently well-known.

The infra-red spectra of some Scheelite structures $CaWO_4$, $CaMoO_4$, $PbWO_4$ and $PbMoO_4$ were recorded and discussed by Khanna and Lippincott [88].

Hezel and Ross [89] gave an extensive discussion on the occurrence of forbidden transitions in the infra-red spectra of tetrahedral anions of the type MXO_4, where $M =$ Cl, S and P, referring to 61 papers.

Müller [90] discusses the factor-group analysis of potassium perrhenate $KReO_4$, using previously-published infra-red and Raman spectra. The space group is C_{4h}^6, where site symmetry of the ReO_4^- ion is S_4; there are only two formula units per spectroscopic unit cell, and all eight infra-red active modes are found.

Scheuermann, Ritter and Schutte [91] discussed the spectra of $SrCrO_4$ and $PbCrO_4$; these salts are monolinic and isomorphous, but have remarkably different spectra as inversion doubling occurs in $SrCrO_4$, but not in $PbCrO_4$. It was also found that the Raman spectra of the orthorhombic and monoclinic $PbCrO_4$ are identical.

d) Tetrahedral XH_4 Structures

The tetrahedral structures $XH_4^\pm$ are still extensively investigated, and the factor-group splittings reported, although the attention is now also focussed on the single-crystal spectra.

Durig, Antion and Pate [92] recorded the IR spectra of thin films of phosphonium bromide PH_4Br and PD_4Br at $-170\,^\circ C$ (the salt crystallizes with space group D_{4h}^7-P4/nmm, $Z = 2$) as well as those of the isomorphous PH_4I [93]; the librations and translations were assigned. The barrier opposing the rotation of the PH_4^+ and PD_4^+ ions in the bromide is 7.8 kcal/mole ($\nu = 354$ cm^{-1}) and 7.9 kcal/mole ($\nu = 254$ cm^{-1}), which is higher than the barrier calculated from NMR-data (7.03 kcal/mole) by Tsang, Farrar and Rush [94]. The behaviour of the phosphonium salts is quite analogous to that of the NH_4^+ salts in broad detail. The laser Raman spectra of PH_4I, PH_4Br and PH_4Cl have been measured by Rush, Melveger and Lippincott [95] at 25 °C for the iodide and bromide and at ~ 5 °C for the chloride, the factor-group-assignments made, and rotational barriers calculated (and compared with those of the ammonium halides); they also give a brief discussion of the phase relationships of the phosphonium halides and the ammonium halides. A neutron diffraction study of PH_4Br is reported by Schröder and Rush [96].

The ammonium halides have again been subjected to a close scrutiny by several authors, confirming the exhaustive investigations of Hornig and Co-workers in the early fifties [97-100]. Perry and Lowndes [101] studied the external optical phonons of all the ammonium halides, also giving a useful table summarizing the known phase changes; the results confirm the factor group predictions. Durig and Antion [102] measured the low-frequency vibrations of NH_4I and NH_4Br from -170 °C to room temperature; the Raman spectrum of phase I confirmed the Whalley and Bertie [103] theory of transitions in optically-disordered lattices, and the phase changes are extensively discussed. Schumaker and Garland [104] made an investigation into the structural and ordering changes in the infra-red and Raman spectra of NH_4Cl and NH_4Br and their fully-deuterated analogues. They recorded the spectra of thin films on alkali-halide plates at numerous temperatures, especially near the transition points and determined the intensities; this method was also used by Schutte and Heyns [105] in their study on $(NH_4)_2SO_4$ and applied to other compounds. Schumaker and Garland [104] give tables of the fundamental frequencies of the ammonium halides (excepting NH_4F) in the various recorded phases, including the fundamental modes, ν_6 (the torsion) and the two lattice modes ν_5 (TO) and ν_5 (CO), as well as the overtones. Of especial importance is the temperature-variation of the first overtone of the ν_6-librational mode in NH_4Br (the detailed behaviour of this band in NH_4Cl has been reported previously [106]). The temperature-dependence of all the main bands, as well as the overtones are also given, and show details of the transitions. Schumaker and Garland [104] conclude that infra-red spectroscopy can provide significant information on the phase behaviour of cooperative systems, but that further work is needed on the theoretical aspects of phase changes and selection rules, especially of the overtones; these conclusions also follow from the work of Schutte and Heyns [105], where small changes in lattice, are picked up by infra-red techniques (see Paragraph 7 of this review). Phase studies were also reproted for NH_4Cl by Bartis [107], while the neutron-determined barriers of the NH_4^+ ion were discussed by Leung, Taylor and Havens [108] in a paper on the phase transitions of ammonium halides as determined by neutron-scattering. Pistorius [109] reported the high-pressure phase diagram of the ammonium halides, while Rapoport and Pistorius [110] reported the phase diagram of NH_4NO_3.

Goldfinger and Verhaegen [111] determined the stability of the gaseous ND_4Cl molecule using a Knudsen cell and a quadrupole mass spectrometer. The main decomposition products were ND_3 and DCl, but small proportions of ND_4Cl-molecules were also observed. These are 'ionic molecules' $NH_4^+Cl^-$ of the type predicted by Clementi [112,113] and by Mulliken [114], but occur at small concentrations — Goldfinger and Verhaegen state that the mass-spectrometric technique was probably the only technique which could identify these ionpairs. However, Sterling and Haines [115] have published a spectrum showing NH_4Cl in the gas phase at 180 °C; NH_3 and HCl are present, as well as $NH_4^+Cl^-(g)$ which has a spectrum remarkably close to that of the high-temperature crystalline modification, although Walker [116] has failed to confirm this. A re-examination of the alkali halides in the gas phase is currently under way in this laboratory, because the spectrum of NH_4Cl [115] indicate bands at ca. $100-200$ cm^{-1} lower than in the solid state stretching region.

Rather surprising, the spectrum of NH_4F has not been reinvestigated after the study of Plumb and Hornig [117] who determined that NH_4^+ lies on a T_d site and the nearest-neigbour F^- ions are indeed tetrahedrally arranged. Perhaps the recent determination of the phase diagram of NH_4F by Kuriakose and Whalley [118] will lead to new spectroscopic research, especially in the liquid region; four phases are reported in the range 1 to 20 kbar and 20° to 340 °C. Calvert and Whalley [120] have determined the structure of NH_4F IV as cubic O_h^6-Fm3m (NaCl-structure) with a (probably) disordered NH_4^+-structure.

The infra-red spectrum of polycrystalline NH_4CN was reported by Clutter and Thompson [119] for temperatures between 125° and 4 °K. The H-bond in NH_4CN seems to be weaker than that of NH_4Cl. The CN group is orientationally indistinguishable between CN and NC, so that the space group is D_{4h}^{10}-P4$_2$/mcm, $Z = 2$; if the CN$^-$ are all pointing in one direction the space group would be C_{2v}^4-Pma2. The resulting spectra are so simple that it seems as if the indistinguishability between the CN-orientations is of little consequence, and the D_{4h}-factor group indicates most of the observed features.

68

The torsion is not observed, but the combination mode is found, giving $\nu_6 \sim 410$ cm^{-1} for NH$_4^+$. The simple spectrum found seems very similar to the surprisingly simple spectrum found by Plumb and Hornig [117] for NH$_4$F with its very strong hydrogen bonds and a librational mode of 523 cm^{-1} for NH$_4^+$; the ν_6 of NH$_4$CN, although not exactly known, is higher than that of all the other halides of NH$_4^+$ (see Schumaker and Garland [104]).

e) Various Ions and Molecules

The azide ion N$_3^-$: Bryant [121] has made a study of the infra-red spectrum of CsN$_3$ over the frequency range 30–4000 cm^{-1}, using the Bhagavantam and Venkaterayudu method [54] to determine the splittings; the spectra were recorded at 77 °K and at 298 °K. The symmetric stretching frequency should not occur (it is symmetry-prohibited); its occurence is explained by the coupling of lattice modes observed in the far-infrared part of the spectrum with the B_{2g}-species of ν_1. Assignments are made to ~50 absorptions, confirming the reported crystal structure D_{4h}^{18}-I4mcm. α-lead azide, as reported in the neutron-diffraction study by Choi and Boutin [122], has the space group Pcmn and show *four* different N$_3^-$ groups in the structure; the four crystallographically different N$_3^-$ groups are all essentially linear, but each one is in a different environment. The spectrum of α-Pb(N$_3$)$_2$ should probably be worth-while to investigate (it should be remembered that azides are explosive!). Iqbal, Brown and Mitra [123] report a very interesting analysis of the single crystal vibrational spectrum of barium azide Ba(N$_3$)$_2$; the infrared and Raman spectra confirm the presence of an asymmetric azide ion in the unit cell (type II-ions in the C_{2h}^2-P2$_1$/m, $Z = 2$ unit cell, as determined by Choi [124]).

Phosphorus P_n: Durig and Casper [125] analyzed the infra-red and Raman spectra of amorphous solid red phosphorus, and concluded that the structure is polymeric with either a puckered 4-membered ring (D_{2d}) or two equilateral triangles with a common base (C_{2v} symmetry).

Amide ion NH$_2^-$: Bouclier, Portier and Turrell [126] gave a factor-group analysis of both phases (α, tetragonal D_{4h}^{17}-I4$_1$/amd with $Z = 4$; β disordered cubic of the NaCl type O_h^5-Fm3M with $Z = 2$). The unit cell contains only four coupling NH$_2^-$-groups, because for the I-groups the Bravais unit cell (spectroscopic unit cell) is half the crystallographic unit cell, so that their modes are doubled. If the β-phase is cubic, as their X-ray results seem to indicate with $Z = 2$, it means that the Bravais cell contains only $^1/_2$ Ca(NH$_2$)$_2$ — all F-groups have Bravais cells which are $^1/_4$ of the crystallographic unit cell — and hence no factor group splitting should occur, contrary to that observed; it seems that further work is needed to unravel the spectrum of both phases satisfactorily, *e.g.* a neutron-diffraction investigation to confirm the statistical disorder which Bouclier, Portier and Turrell [126] postulate to account for the cubic phase β structure. The NH_2^- ion seems to enjoy a reasonable freedom of movement in the lattice of $Ca(NH_2)_2$ and $Sr(NH_2)_2$, because the temperature dependence of the wide-line N.M.R. spectrum shows evidence of both librating motion, as well as rotations through $180°$ about its C_2-axis, according to Dufourcq, Chézeau and Lemanceau [127].

Substituted NH_4^+ ions: Whalley [128] discussed the spectroscopic effects of orientional disorder about one axis (in contrast to the disorder about three axes as described by Whalley and Bertie [103] and Bertie and Whalley [129] in the α-phases of the methylammonium halides. In principle, all vibrations of an orientational disordered crystal are spectroscopically active, but if the disorder is only about one axis, some restrictions operate, the symmetric bands are sharp in the one-dimensional disordered case, but the bands due to asymmetric vibrations (E) are broad. Whalley use the infra-red results of Sandorfy *et al.* [130,131] of the CH_3-ammonium halides to illustrate the effect which is predicted from interionic coupling of the E-modes. No such effect is visible in the spectrum of the methoxyammonium ion $CH_3ONH_3^+$ reported by Nelson [132].

Water of hydration: Berthold and Weiss [133] used infra-red spectroscopy to help solve the crystal structures of the hydrates $Na_2S_2O_6.2H_2O$ and $Li_2S_2O_6.2H_2O$ so that a complete atomic arrangement was obtained.

Hohler and Lutz [134] made a study of the combinational modes of gypsum $CaSO_4.2H_2O$ in the region $1200-10,000$ cm^{-1}, using polarized light; the IR spectrum shows a great number of bands, the most of which are assigned and the rôle of the protons in the lattice is discussed (see also Hass and Sutherland [135] and Schaack [136,137]. Beattie, Gall and Ozin [138] analyzed the vibrational spectrum of $Na_2S_2O_6.2H_2O$ (dithionate); the factorgroup analysis is described in detail because it is particularly easy to visualise.

α-Quartz: Merten [139] measured the directional dispersion of the extraordinary infra-red bands of the *n-*, *k-* and the reflection spectra of α-quartz; it is shown that resonances previously thought to be combinational modes are wrongly assigned.

Rhombic Sulphur: Ozin [140] has obtained single-crystal polarization Raman data for the rhombic S_8 solid with its D_{4d}-molecules (D_{4h} in the gas phase) and used the factor group method to assign all the observed modes; he suggests that the low frequency IR modes were correctly assigned by Scott, McCullough and Kruse [141]. The infra-red data of Chantry, Anderson and Gebbie [142] were also utilized in the assignment.

The H-bond: Maréchal and Witkowski [143] gave a theoretical approach in order to describe the peculiar vibrational features of H-bonded crystals; they derive a Hamiltonian to describe the vibrations of a linear crystal (the X-H modes in $X\text{-}H\cdots X$), but no numerical results are derived.

3.2. Organic Compounds

Very many organic compounds have been studied in their solid states, but the problem is very often to obtain suitable crystals for a polarized light study. Kruse [144] has built an automatic crystal growth apparatus in which thin pure single crystals of organic compounds can be grown *along any pre-determined direction* between alkali-halide plates spaced *ca.* 25 μ apart. He applied this to the crystals of *p*-dichloro-benzene, a compound which has caused much confusion before; the quality of the dichroic spectra is excellent, and enabled him to satisfactorily assign the vibrations. Suzuki and Ito [145] also reported the polarized spectra of *p*-dichloro- and *p*-dibromo-benzene. Benzene has been studied by Marzocchi, Bonadeo and Taddei [146], by Greer, *et al.* [147] and the normal modes of crystalline benzene and naphtalene by Harada and Shimanouchi [148] (see also Greer *et al.* [147]), while the Raman spectrum of C_6H_6 was studied by Gee and Robinson [149].

Naphtalene and anthracene were studied using Raman spectroscopy by Suzuki, Yokohama and by Ito [150]; the normal selection rules are not applicable to the Raman spectrum of single crystals of anthracene, as discussed by Ting [151]. Anthracene was also studied by Bree and Kydd [152] and anthracene-d_{10} by Bree and Kydd [153].

Fluorene was studied by Witt [154] and by Bree and Zwarich [155].

The polarized IR spectrum of single crystals of phenantrene was analysed by Schettino, Neto and Califano [156], that of 9, 10-anthraquinone by Pecile and Lunelli [157], pyrimidine by Brana, Adembri and Califano [158], and that of CH_2Br_2 by Marzocchi *et al.* [159] and by Brown *et al.* [160]. Furan and pyrrole were studied by Loisiel and Lorenzelli [161], cyclopropyne by Jubino, Dellepiane and Zerbi [162], and cyclohexane by Obremski, Brown and Lippincott [163], and cyclopropane by Bates, Sands and Smith [164].

Witt and Mecke [165] compared the orientated crystal spectra of phenantrene and 9–10 dichlorophenantrene, and Chafik and Mecke [170] measured the dichroism of anthracene-d_2 (9–10) from 400–3100 cm^{-1}; the non-planar A_u vibrations are discussed because they appear in spite of the fact that they are prohibited by the site symmetry D_{2h} of the molecule.

Formic acid H.COOH and acetic acid CH_3.COOH were studied in the crystalline state by Carlson, Witkowski and Fateley [166] in the far infra-red region, while Fukushima and Zwolinski [167] gave a normal coordinate calculation of acetic acid dimers $(CH_3.COOH)_2$; the acetate ion in $Li(CH_3.COO).2H_2O$ was studied by Cadene [168], while DiLauro, Califano and Adembri [169] studied the crystal spectra and normal modes of the anhydrides of maleic and succinic acids in polarized light.

Rey-Lafon *et al.* [171] analyzed the solid state vibrational spectrum of 1, 3, 5-trinitro hexahydro-s-triazine, and show that coupling between the NO_2-groups occur.

4. Solid-Solution and Matrix-Isolation Spectra

Not very many solid-solution studies were reported, mainly because all the "easy" substituted ions have been previously experimentally treated. The tetrahedral ions continue to be studied, and BD_4^- in alkali halides was studied by Coker and Hofer [172]; see also Heyns, Schutte and Scheuermann [214]. The solid solutions were prepared by diffusing the deuteride into optical grade crystals, and the IR spectra were recorded at 90 °K, yielding all 24 vibrational bands of BH_4^-, BH_3D^-, $BH_2D_2^-$, BD_3H^- and BD_4^-. The trends are similar to those first found by Schutte and Ketelaar for BH_4^- solid solutions [173]. There is still no satisfactory theory for the spectra of ions in solid solutions, but Metselaar [174] gave a theory of lattice vibrations of solid solutions in his thesis on the NO_3^- ions in alkali halides.

Bonn, Metselaar and Van der Elsken [175] discussed the spectra of NO_3^- in various alkali-halide lattices in terms of rotatory motions of the nitrate groups, but Kato and Rolfe [176] have questioned this approach, and prefer isotopic interaction instead.

Manzelli and Taddei [177] succeeded in substituting the MnO_4^- ion on lattice sites in KCl, RbCl, KBr and RbBr, and the solid solution spectra are superposed upon the $KMnO_4$ spectrum; tables of frequencies are given.

Krynauw and Schutte [178-181] in a series of papers discussed the infrared spectra of solid solutions of ClO_4^- and ClO_3^- in alkali-halide lattices, measured the absolute absorption intensities and discussed the lattice-dependence of the ν_3 and ν_4-modes of the ClO_4^- ion in these lattices.

McKean [182] considered the matrix shifts and lattice contributions from a classical electrostatic point of view, using a multipole expansion of the electrostatic energy to represent the vibrating molecule and applied this to the XY_4 molecules trapped in noble-gas matrices. Mann and Horrocks [183] discussed the environmental effects on the IR frequencies of polyatomic molecules, using the Buckingham potential [184], and applied it to HCN in various liquid solvents. Decius [185] analyzed the problem of dipolar vibrational coupling in crystals composed of molecules or molecular ions, and applied the derived theory to anisotropic Bravais lattices; the case of calcite (which introduces extra complications) is treated separately. Freedman, Shalom and Kimel [186] discussed the problem of the rotation-translation levels of a tetrahedral molecule in an octahedral cell.

The technique of matrix isolation continued to be exploited extensively in order to isolate species which cannot be isolated in another way, *e.g.* $CH_3.COOH$ monomers by Berney, Redington and Lin [187], glycine by Grenie, Lasseques and Garringon-Lagrange [188], group IV oxides by Ogden and Ricks [189], H_2S by Tursi and Nixon [190] and by Pacansky and V. Calder [191], SO_2 by Allavena *et al.* [192], the dimer of nitric oxide $(NO)_2$ by Guillory and Hunter [193] who suggest *cis* and *trans* O=N=N=O), a series of tetrahedral molecules by King [194], the hydrogen dichloride ion $ClHCl^-$ by Milligan and Jacox [195], the SH radical by Acquista and Schoen [196], and HCl by Keyser and Robinson [197] and by Von Holle and Robinson [198].

5. Attenuated Total Reflection

Attenuated total reflection or frustrated total reflection, is being increasingly used to obtain solid state spectra, and to reveal solid state surface effects (see Paragraph 3.1 above); the general technique is described by Davies [199]; see also Wilks and Hirshfeld [19]. Brooker [200] reported the detection of transverse modes in $NaNO_2$, $NaNO_3$ and $CaCO_3$ (calcite) using this technique. Devlin, Pollard and Frech [201] made a detailed study of the ATR IR spectra of uniaxial nitrate single crystals, and it is shown that the band structures are in accord with prediction based upon the damped-oscillator model and the complex form of Snell's law and Fresnel's reflection equations as developed for the uniaxial crystal in question. Ahlijah and Mooney [202] compare the ATR frequencies of solid nitrates (Na, K, NH_4, Mg, Ca, Pb, Th), nitrites (Na, K), NaCN, NaCNS, etc., with the solution ATR spectra; this method can also be used for quantitative analyses. Deane, Richards and Stephen [203] used the ATR method to determine the bond orientations in uranyl nitrate hexahydrate $UO_2NO_3.6H_2O$.

This is a technique which deserves to be more fully exploited.

6. Pressure Effects

The infra-red spectrum of a compound is also pressure-dependent, but the experimental detection is often very difficult as high pressures are involved. Lattice modes are especially sensitive, as the recent studies of cubic ionic crystals of Ferraro, Postmus and Mitra [204], Ferraro et al. [205,206] show. Ferraro [207] studied the internal and lattice modes of K_2PtCl_4, K_2PdCl_4, K_2PtCl_6 and K_2PdCl_6 under pressure and concluded that the lattice modes show a higher pressure dependency than the internal modes. Blinc, Ferraro and Postmus [208] determined the pressure effects of the far-IR spectra of paraelectric KH_2PO_4 and RbH_2PO_4; the room-temperature spectra of RbH_2PO_4 show the presence of a new phase with ordered protons. The Raman spectrum of α-quartz at high pressures was investigated by Asell and Nicol [209] up to 40 kbar, and the pressure dependence of the lines reported.

Fong and Nichol [210] made a very interesting high-pressure (up to 40 kbar) Raman spectroscopic study of $CaCO_3$ II and $CaCO_3$ III. They report 13 lines for phase II and 20 for III. This study disproves the earlier suggestions [211-213] that $CaCO_3$ II and KNO_3 III and that $CaCO_3$ III and KNO_3 II may be isostructural, and alternative structures for these $CaCO_3$ phases are considered. The frequency of the Raman active phonon at $150-175$ cm^{-1} is also very much dependent upon the applied pressure, being the lowest in phase II and highest in phase III; in fact, the phase transitions: room pressure phase → phase II → phase III could be determined. Phase II is also characterized by the splitting of the ν_4 $CO_3^=$ mode into a pair of lines at 715 and 721 cm^{-1}.

7. Phase Changes

The ease with which IR-spectroscopy detects the phase changes of ammonium halides was already referred to in Paragraph 3.1d. The technique was used by Schutte and Heyns [105] to determine the phase changes in $(NH_4)_2SO_4$; they also found an IR-NMR line-width correlation at 163 °K, the first time that such a correlation has been established. There is no evidence that in $(NH_4)_2SO_2$ there is any "free rotation" of NH_4^+, since the torsional mode was identified. Heyns, Schutte and Scheuermann [214] also used this technique to establish the phase transition in $NaBD_4$, which should be analogous to that found in $NaBH_4$ by Ketelaar and Schutte [215] and by Schutte [216]; the phase transition was found to occur at $197 \pm 1°$ using both Raman and IR-spectroscopy, in agreement with the Na^{23} resonance studies of Subratová [217]. The ν_2-Raman components are especially temperature-dependent, and offer an excellent opportunity to determine the phase transition because there are both a frequency shift and a reduction in the number of components. The spectra of NH_4CuCl_3 (ammonium trichlorocuprate) should show evidence of severe distortion according to the X-ray results of Willett et al. [218], although Heyns and Schutte [219] do not find any evidence of this and bands which should be activated by both the site and factor group perturbations are not observed; it is, therefore, concluded that the NH_4^+-ion must have considerable rotational freedom. A possible antiferromagnetic transition was also observed by these authors. Schutte and Van Rensburg [220] studied the temperature behaviour of NH_4HSO_4 and determined the transition temperature. NH_4BF_4 was studied by Caron et al. [221]. Schutte and Heyns [222] also studied $(NH_4)_2Cr_2O_7$, and observed three phase transitions and in the deuterated salt there are two phase changes; the behaviour of NH_4^+ with respect to reorientation is also discussed.

Castellucci, Sbrana and Verderame [223] report a phase transition at $-60°$ in C_5H_5N (deutero pyridine), but could not induce a similar transition in normal pyridine (see also Loisel and Lorenzelli [161]). May and Pace [224] report two phases in CH_3SH and the deuterated CH_3SD, using IR-techniques, as well as X-ray diffraction; the transition takes place at ~ 136 °K. Brüesch and Günthard [225] report phase transitions for bicyclo (2.2.2) octane, triethylenediamine and quinuclidine.

Craft and Slutsky [226] report an ultrasonic and IR-study of the λ-point transition in $NaNO_3$ at 276.7 °C, where the appearance of ν_1 of NO_3^- below the transition point is used as a phase-change criterion. Sato [227] report the temperature-variation of the vibrational bands of KH_2PO_4 and KD_2PO_4 to determine the well-known phase transition; this throws some light on the nature of the phase transition.

Wiener, Levin and Pelah [228] studied the proton dynamics in KH_2PO_4-type ferro-electrics by means of IR-absorption, and observed spectral changes when traversing the Curie temperature. The H-bonds are discussed and the correlation tables are given. The same authors [229] studied the anti-ferro-electric transitions in $NH_4H_2PO_4$ and in $NH_4H_2AsO_4$, also using IR-absorption spectra; this throws some light on the structures of the phases.

8. Complexes

The IR-spectra of very many — perhaps most — new inorganic coordinate complexes are yearly recorded. Most of these spectra are just fingerprinting and identification purposes or for empirical correlations and predictions. There is a definite need to put most of these empirical correlations — some of the evidence is usually very flimsy, spectroscopically speaking, and the reader is warned not to put too much faith in it — on a firmer footing through normal-coordinate analyses. Only a few solid state compounds were described by such theoretical techniques in the past few years, the most important being listed below.

Jones, who has been very active in the description of the complex cyanides, has published a short paper on $Cs_2LiCo(CN)_6$, together with Swanson [230]; the new observations on the cobalticyanide ion show that the old assignments of the C—Co—C deformation vibration are in error for the potassium salt (see Jones [231,232], Bloor [233] and McAllister [234]). The Raman laser spectrum of $K_2Ag(CN)_2$ was reported by Loehr and Long [235]; this spectrum was exceedingly thoroughly investigated by Jones [236-238] by IR and by Bottger [239] using far infra-red radiation, and the Raman evidence is in favour of the attachment of CN through the C-atom to the metal. The $Ag(CN)_2^-$ ion in the K, Na and Tl salts was also studied by Chadwick and Frankiss [240]. The vibrational spectra of many hexacyanides $M(CN)_6^{x-}$, M = Co, Fe, Mn, Cr, Ru have been remeasured in the solid state by Griffith and Turner [241], and analysed with reference to 31 literature references.

Krasser [242] made a normal mode calculation of the vibrational modes of $Na_2(Fe(CN)_6NO)$ which gives a C_{4v} anion.

The octacoordinated ions $Mo(CN)_8^{4-}$, $W(CN)_8^{4-}$ and TaF_8^{3-} were studied by Raman and by infra-red spectra in solution and as the potassium dihydrate salts by Hartman and Miller [243]. The solution spectra of the cyanides are consistent with the square Archimedian antiprism structure which belongs to the point group D_{4d}, but do not establish a proof. The solid state spectra corroborate the D_{2d}-structure established for $K_4Mo(CN)_8.2H_2O$ by Hoard and Nordsieck [244], although the ion in Na_3TaF_8 has the D_{4d}-structure as established by Hoard et al. [245]; this study thus clears up the discrepancy between the work of Hoard and Nordsieck [244] and the solution infrared and Raman work of Stammreich and Scala [246,247].

Siebert and Eysel [248] examined the Raman and infra-red spectra of the amine complexes $(Co(NH_3)_6^{3+}, Co(ND_3)_6^{3+}$ and $Cr(NH_3)_6^{3+}$ with various anions, and the influence of the lattice symmetry upon the vibrations is discussed.

Nakagawa and Shimanouchi [249], discussed the far-IR and lattice spectra of $Co(NO_2)_6^{3-}$ salts, giving a complete vibrational analysis. This paper is the ninth by the same authors on the theoretical description of coordination compounds (see References in [249]). The same type of calculation is reported for the structure $Co((NO_2)_n(NH_3)_{6-n})^{(3-n)+}$ by the same authors [250].

Debeau and Poulet [251] reported an interesting study of 31 crystalline hexachloro MCl_6 and hexabromo MBr_6 metallates and the force constants are calculated with a modified Urey-Bradley force field; the relation between the metal-X stretching frequency and the MX covalency is discussed.

Hartley and Ware [252] reported Raman and IR data for solid ferrocenes and the isoelectronic cation $(Co(C_5H_5)_2^+)$; Long and Huege [253] also reported the Raman spectrum of ferrocene, and the two investigations are in substantial agreement. This work seems to settle the assignment problem in these compounds. Bunker [254] used the theory of non-rigid vibrations developed empirically by Longuet-Higgins [255], and predicted five coincident bands in the Raman and IR, a fact not predicted by the normal selection rules.

Murrell [256] shows theoretically that for MX_6 compounds the three bond stretching force constants can be calculated sufficiently accurately from IR and Raman spectra; from this the complete harmonic force field can be deduced. This is an important simplification, and bond strengths can be more conveniently calculated. Murrell shows that e.g. in WF_5Cl that replacement of F by Cl strengthens the *trans* WF bond and weakens the *cis* WF bonds. Lane and Sharp [257] discuss the hexafluorometallates of Group IV, using the factor group approach, and Adams and Newton [258] report the single crystal IR and Raman spectra of K_2PtCl_4.

Beattie et al. [259] analysed the factor group splittings (by Raman and IR-spectroscopy) of β-NiF_2, $SbCl_4F$, VF_5, NbF_5 and TaF_5. Buttery et al. [260] analysed the crystal spectrum in the CO stretching region of π-$C_6H_6Cr(CO)_3$, giving also a force-constant calculation; they followed this study up with one on the methyl-substituted compound [261].

9. Solidified Gases

Solidified gases are still studied, and the hydrogen halides, because of their simple vibrational structure, are favourites. Examples are HF by Kittelberger and Hornig [262], HCl by Wang and Fleury [263] and by Friedrich and Carlson [264], all the hydrogen halides by Savoie and Anderson [265] and the Raman spectra by Ito, Suzuki and Yokoyama [266]. De Bettignies and Wallert [267] gave an interesting report on the variations in the Raman spectrum of NH_3 with temperature, starting from the solid state. Herzog and Schwab [268] looked at the Raman spectra of liquid and solid NO, observed no solid Raman lines, but a few liquid lines, and use quoted literature IR values to show that *cis* and *trans* forms are present (see also Guillory and Hunter [193]).

Johansen [269] calculated the isotopic frequency shifts occurring in the molecular crystals of carbon monoxide CO, using the infra-red spectrum of Ewing and Pimentel [270]; the crystal is considered to be an ensemble of harmonic oscillators and the FG-method is used.

10. Conclusion

This review shows that infra-red spectroscopy is used very much today to unravel the nature of the solid state and the forces operating in crystals. It is clear that the advent of the automatic recording laser Raman spectrometer has enhanced the usefulness of the infra-red spectrometer for solid-state studies. There is still scope for much work, *e.g.* in the study of phase changes, and especially in the structure of complexes in order to clear up many of the problems found in the empirical interpretation of the spectra.

References

[1] Schutte, C. J. H.: The ab-initio calculation of molecular vibrational, frequencies and force constants. Struct. Bonding 9, 213 (1971).

[2] Schutte, C. J. H.: The adiabatic Born-Oppenheimer approximation. Quart. Rev. (London) 25, 393 (1971).

[3] Wolniewicz, L.: J. Chem. Phys. 45, 515 (1966).

[4] Bak, B., Clementi, E., Kortzeborn, R. N.: J. Chem. Phys. 52, 764 (1970).

[5a] Colon, R., et al.: J. Phys. Radium 15, 641 (1954).

[5b] This effect is also discussed by Ketelaar, J. A. A.: Spectrochim. Acta 15, 237 (1959). – The theory of this effect is further discussed by Bratož, S., Martin, M. L.: J. Chem. Phys. 42, 1051 (1965) for mixtures of HCl + X, where X is an inert gas,

C. J. H. Schutte

N_2, CO etc. See also Van Kreveld, M. E., Van Aalst, R. M., Van der Elsken, J.:
J. Chem. Phys. *55*, 2853 (1971).

5c) Ketelaar, J. A. A., Hooge, F. N.: J. Chem. Phys. *23*, 749 (1955).

6) Witkowski, A.: J. Chem. Phys. *52*, 4403 (1970).

7) Halford, R. S.: J. Chem. Phys. *14*, 8 (1946).

8) Couture, L.: J. Chem. Phys. *15*, 153 (1946).

9) Hornig, D. F.: J. Chem. Phys. *16*, 1063 (1948).

10) Carter, R. L.: J. Chem. Educ. *48*, 297 (1971).

11) Fateley, W. G., McDevitt, N. T., Bentley, F. M.: Appl. Spectry *25*, 155 (1971).

12) Bertie, J. E., Bell, J..W.: J. Chem. Phys. *54*, 160 (1971). – See also Bertie, J. E.,
Kopelman, R.: J. Chem. Phys. *55*, 3613 (1971) for a more clarifying treatment.

13) Herzberg, G.: Molecular spectra and molecular structure. In: Infrared and Raman
spectra of polyatomic molecules, Vol. II. Princeton: D. Van Nostrand Company
Inc. 1945.

14) Henry, N. F. M., Lonsdale, K. (eds.): International tables for X-ray crystallography, 2nd edit. Vol. 1. Symmetry groups. The International Union for Crystallography. Birmingham: Kynoch Press 1965.

15) Adams, D. M., Newton, D. C.: J. Chem. Soc. *1970A*, 2822.

16) Adams, D. M., Newton, D. C.: Tables for factor-group analysis. Croydon: Beckman-
RIIC Ltd. 1970.

17) Vedder, W., Hornig, D .F., in:Advances in spectroscopy (ed. H. W. Thompson),
Vol II. London: Interscience Publishers Inc. 1961.

18) Mitra, S. S., in: Solid state physics (eds. F. Seitz and D. Turnbull), Vol. 13, 1962.

19) Wilks, P. A., Jr., Hirshfeld, T., in: Applied spectroscopy reviews (ed. E. G. Brame,
Jt.). Vol. 1. New York: Marcel Dekker Inc. 1968.

20) Orville-Thomas, W. J.: J. Mol. Struct. *1*, 357 (1967/1968).

21) Eyring, H., Christiansen, C. J., Johnston, H. S. (eds.): Annual reviews of physical
chemistry, Vol. 1. Palo Alto: Annual Reviews Inc. 1950.

22) Buijs, K., Schutte, C. J. H.: Spectrochim. Acta *18*, 307 (1962). – Spectrochim.
Acta *17*, 927 (1961) (Simple inorganic carbonates).

23) Schroeder, R. A., Weir, C. E., Lippincott, E. R.: J. Res. Nat. Bur. Std. *66A*, 407
(1962) (Low-temp. IR of carbonates and nitrates for rotational barriers.)

24) Weil, W. A.: Advan. Chem. Ser. *33*, 72 (1961).

25) Cooney, R. P. J., Thayer, C., Li, P. C., Devlin, J. P.: J. Chem. Phys. *51*, 302 (1969).

26) Devlin, J. P., James, D. W., Frech, R.: J. Chem. Phys. *53*, 4394 (1970).

27) James, D. W., Leony, W. H.: J. Chem. Phys. *51*, 640 (1969).

28) Brooker, M. H.: J. Chem. Phys. *53*, 2670 (1970).

29) Schutte, C. J. H.: Z. Krist. *126*, 397 (1968).

30) Birnstock, R.: Z. Krist. *124*, 310 (1967).

31) Brooker, M. H., Irish, D. E., Boyd, G. E.: J. Chem. Phys. *53*, 1083 (1970).

32) Schutte, C. J. H.: Z. Phys. Chem. N.F. *39*, 241 (1963).

33) Sirdeshmukh, D. B.: J. Phys. Chem. Solids *27*, 1157 (1966).

34) Belyustin, A. V., Levina, I. M., Novolsel'tseva, T. P.: Sov. Phys. Cryst. *13*, 633
(1969).

35) Bon, A. M., Vergnoux, A. M.: Compt. Rend. *271B*, 245 (1970).

36) James, D. W., Leony, W. H.: J. Chem. Phys. *49*, 5089 (1968).

37) Miller, R. E., Getty, R. R., Treuil, K. L., LeRoi, G .E.: J. Chem. Phys. *51*, 1385
(1969).

38) Nakayawa, I., Walter, J. L.: J. Chem. Phys. *51*, 1389 (1969).

39) Rosseau, D. L., Miller, R. E., LeRoi, G. E.: J. Chem. Phys. *48*, 3409 (1968).

40) Porto, S. P. S., Giordmaine, J. A., Damen, T. C.: Phys. Rev. *147*, 608 (1966).
41) Hellwege, K. H., Lesch, W., Philal, M., Schaak, G.: Z. Physik *232*, 61 (1970).
42) Decius, J. C., Malan, O. G., Thompson, H. W.: Proc. Roy. Soc. (London) *275A*, 295 (1963).
43) Sterzel, W., Chlorinsky, E.: Spectrochim. Acta *24A*, 353 (1968).
44) Müller, A., Fadini, A.: Z. Phys. Chem. N.F. *54*, 29 (1967).
45) Becher, H. J., Ballein, K.: Z. Phys. Chem. N.F. *54*, 302 (1967).
46) Pfeiffer, M.: Z. Phys. Chem. N.F. *61*, 253 (1968).
47) Newman, R.: J. Chem. Phys. *20*, 1663 (1952).
48) Ito, K., Bernstein, H. J.: Can. J. Chem. *34*, 170 (1956).
49) Harvey, K. B., Morrow, B. A., Shurvell, H. F.: Can. J. Chem. *41*, 1181 (1963).
50) Schutte, C. J. H., Buijs, K.: Spectrochim. Acta *20*, 187 (1964).
51) Hammaker, R. M., Walters, J. P.: Spectrochim. Acta *20*, 1311 (1964).
52) Donaldson, J. D., Knifton, J. F., Ross, S. D.: Spectrochim. Acta *20*, 847 (1964).
53) Kuroda, Y., Kubo, M.: Spectrochim. Acta *23A*, 2779 (1967).
54) Bhagavantam, S., Venkaterayudu, T.: The theory of groups and its application to physical problems, Walthair: Andra University Press 1951, this method is essentially the same as that of Hornig [9].
55) Buchanan, R. A., Caspers, H. H., Murphy, J.: Appl. Opt. *2*, 1147 (1963).
56) Oehler, O., Günthard, Hs. H.: J. Chem. Phys. *48*, 2036 (1968).
57) Oehler, O., Günthard, Hs. H.: J. Chem. Phys. *48*, 2032 (1968).
58) Mitra, S. S.: J. Chem. Phys. *39*, 3031 (1963).
59) Oehler, O., Günthard, Hs. H.: J. Chem. Phys. *51*, 4714 (1969).
60) Lide, D. R., Kuzkowski, R. L.: J. Chem. Phys. *46*, 4768 (1967).
61) Acquista, N., Abramowitz, S., Lide, D. R.: J. Chem. Phys. *49*, 780 (1968).
62) Acquista, N., Abramowitz, S.: J. Chem. Phys. *51*, 2911 (1969).
63) Acquista, N., Schoen, L. J., Lide, D. R., Jr.: J. Chem. Phys. *48*, 1534 (1968).
64) Klein, M. V., Wedding, B., Levine, M. A.: Phys. Rev. *180*, 180 (1969).
65) Devonshire, A. F.: Proc. Roy. Soc. (London) *153A*, 601 (1936).
66) Scott, R. S., Flygare, W. H.: Phys. Rev. *182*, 445 (1969).
67) Shore, H. B.: Phys. Rev. *151*, 570 (1966).
68) Bosshard, U., Dreyfus, R. W., Känzig, W.: Physik Kondensierten Materie *4*, 254 (1965).
69) Decius, J. C., Lilley, S. A.: J. Chem. Phys. *53*, 2124L (1970).
70) Oehler, O., Günthard, Hs. H.: J. Chem. Phys. *51*, 4719 (1969).
71) Tarte, P.: Spectrochim. Acta *18*, 467 (1962).
72) Tarte, P.: Spectrochim. Acta *19*, 25, 49 (1963).
73) Tarte, P.: Spectrochim. Acta *23A*, 2127 (1967).
74) Carreira, L. A., Maroni, V. A., Swaine, J. W., Jr., Plumb, R. C.: J. Chem. Phys. *45*, 2216 (1968).
75) Griffith, W. P.: J. Chem. Soc. *1969A*, 1372.
76) Ratajczak, H.: J. Mol. Struct. *3*, 27 (1969).
77) Coignak, J.-P.: Compt. Rend. *271B*, 583 (1970).
78) Coignak, J.-P.: Compt. Rend. *271B*, 648 (1970).
79) Hill, R. M., Ichiki, S. K.: J. Chem. Phys. *48*, 838 (1968).
80) Kravitz, C. C., Kingsley, J. D., Elkin, E. L.: J. Chem. Phys. *49*, 4600 (1969).
81) Kingsley, J. D., Mahan, G. D., Kravitz, L. C.: J. Chem. Phys. *49*, 4610 (1969).
82) Davydov, A. S.: J. Exptl. Theoret. Phys. (USSR) *18*, 210 (1948).
83) Davydov, A. S.: Theory of molecular excitons. New York: Book Co. McGraw-Hill 1962.
84) Ananthanarayanan, V.: J. Chem. Phys. *48*, 573 (1968).

C. J. H. Schutte

85) Hester, R. E., Krishnan, K.: J. Chem. Phys. *49*, 4356 (1968).
86) Hendra, P. J.: Spectrochim. Acta *24A*, 125 (1968).
87) Doyle, W. P., Kirkpatrick, I.: Spectrochim. Acta *24A*, 1495 (1968), and references therein.
88) Khanna, R. K., Lippincott, E. R.: Spectrochim. Acta *24A*, 905 (1968).
89) Hezel, A., Ross, S. D.: Spectrochim. Acta *22*, 1949 (1966).
90) Müller, A.: Z. Naturforsch. *21A*, 433 (1966).
91) Scheuermann, W., Ritter, G. J., Schutte, C. J. H.: Z. Naturforsch. *25A*, 1856 (1970).
92) Durig, J. R., Antion, D. J., Pate, C. B.: J. Chem. Phys. *51*, 4449 (1969).
93) Durig, J. R., Antion, D. J., Baglin, F. G.: J. Chem. Phys. *49*, 666 (1968).
94) Tsang, T., Farrar, T. C., Rush, J. J.: J. Chem. Phys. *49*, 4403 (1968).
95) Rush, J. J., Melveger, A. J., Lippincott, E. R.: J. Chem. Phys. *51*, 2947 (1969).
96) Schröder, L. W., Rush, J. J.: J. Chem. Phys. *54*, 1968 (1971).
97) Wagner, E. L., Hornig, D. F.: J. Chem. Phys. *18*, 296, 305 (1950).
98) Plumb, R. C., Hornig, D. F.: J. Chem. Phys. *21*, 366 (1953).
99) Vedder, W., Hornig, D. F.: J. Chem. Phys. *35*, 1560 (1961).
100) Wagner, E. L., Hornig, D. F.: J. Chem. Phys. *17*, 105 (1949).
101) Perry, C. H., Lowndes, R. P.: J. Chem. Phys. *51*, 3648 (1969).
102) Durig, J. R., Antion, D. J.: J. Chem. Phys. *51*, 3639 (1969).
103) Whalley, E., Bertie, J. E.: J. Chem. Phys. *46*, 1264 (1967).
104) Schumaker, N. E., Garland, C. W.: J. Chem. Phys. *53*, 392 (1970).
105) Schutte, C. J. H., Heyns, A. M.: J. Chem. Phys. *52*, 864 (1970).
106) Garland, C. W., Schumaker, N. E.: J. Phys. Chem. Solids *28*, 799 (1967).
107) Bartis, F. J.: J. Chem. Phys. *51*, 5176L (1969).
108) Leung, P. S., Taylor, T. I., Havens, W. W., Jr.: J. Chem. Phys. *48*, 4912 (1968).
109) Pistorius, C. W. F. T.: J. Chem. Phys. *50*, 1436 (1969).
110) Rapoport, E., Pistorius, C. W. F. T.: J. Chem. Phys. *44*, 1514 (1966).
111) Goldfinger, P., Verhaegen, G.: J. Chem. Phys. *50*, 1467 (1969).
112) Clementi, E.: J. Chem. Phys. *46*, 3851 (1967).
113) Clementi, E.: J. Chem. Phys. *47*, 2323 (1967).
114) Mulliken, R. S.: Science *157*, 13 (1967).
115) Sterling, K. J., Haines, R. J.: Anal. Chem. *40*, 1395 (1968).
116) Walker, P.: M. Sc. Thesis, University of South Africa, 1971; the high-temperature cell used in this thesis is described in: Sole, M. J., Walker, P.: J. Sci. Instr. *3*, 394 (1970).
117) Plumb, R. C., Hornig, D. F.: J. Chem. Phys. *23*, 947 (1955).
118) Kuriakose, A. K., Whalley, E.: J. Chem. Phys. *48*, 2025 (1968).
119) Clutter, D. R., Thompson, W. E.: J. Chem. Phys. *51*, 153 (1969).
120) Calvert, L. D., Whalley, E.: J. Chem. Phys. *53*, 2151 (1970).
121) Bryant, J. A.: J. Chem. Phys. *45*, 689 (1966), and references therein.
122) Choi, C. S., Boutin, H.: J. Chem. Phys. *48*, 1397 (1968).
123) Iqbal, Z., Brown, C. W., Mitra, S. S.: J. Chem. Phys. *52*, 4867 (1970).
124) Choi, C.: Acta Cryst. *B25*, 2638 (1969).
125) Durig, J. R., Casper, J. M.: J. Mol. Struct. *5*, 351 (1970).
126) Bouclier, P., Portier, J., Turrell, G.: J. Mol. Struct. *4*, 1 (1969).
127) Dufourcq, J., Chézeau, J. M., Lemanceau, B.: J. Mol. Struct. *4*, 15 (1969).
128) Whalley, E.: J. Chem. Phys. *51*, 4040 (1969).
129) Bertie, J. E., Whalley, E.: J. Chem. Phys. *46*, 1271 (1967).
130) Cabana, A., Sandorfy, C.: Spectrochim. Acta *18*, 843 (1962).
131) Théoret, A., Sandorfy, C.: Spectrochim. Acta *23A*, 519 (1967).

132) Nelson, H. M.: J. Chem. Phys. *53*, 1433 (1970).
133) Berthold, I., Weiss, A.: Z. Naturforsch. *22A*, 1440 (1967).
134) Hohler, V., Lutz, H. D.: Z. Naturforsch. *23A*, 708 (1968).
135) Hass, E., Sutherland, M.: Proc. Roy. Soc. (London) *236A*, 467 (1967).
136) Schaack, G.: Physik. Kondensierten Materie *1*, 245 (1963).
137) Schaack, G.: Z. Physik. *176*, 67 (1963).
138) Beattie, I. R., Gall, M. J., Ozin, G. A.: J. Chem. Soc. *1969A*, 1001.
139) Merten, L.: Z. Naturforsch. *23A*, 1183 (1968).
140) Ozin, G.: J. Chem. Soc. *1969A*, 116.
141) Scott, D. W.: McCullough, J. P. M., Kruse, F. H.: J. Mol. Spectry. *13*, 373 (1964).
142) Chantry, G. W., Anderson, A., Gebbie, H. A.: Spectrochim. Acta *20*, 1223 (1964).
143) Maréchal, Y., Witkowski, A.: J. Chim. Phys. *65*, 1279 (1968).
144) Kruse, K. M. M.: Spectrochim. Acta *26A*, 1063 (1970). – J. Phys. E. *3*, 609 (1970).
145) Suzuki, M., Ito, M.: Spectrochim. Acta *25A*, 1017 (1969).
146) Marzocchi, M. P., Bonadeo, H., Taddei, G.: J. Chem. Phys. *53*, 867 (1970).
147) Greer, W. L., Rice, S. A., Jortner, J., Silbey, S.: J. Chem. Phys. *48*, 5667 (1968).
148) Harada, I., Shimanouchi, T.: J. Chem. Phys. *44*, 2016 (1966).
149) Gee, A. R., Robinson, G. W.: J. Chem. Phys. *46*, 4547 (1967).
150) Suzuki, M., Yokohama, T., Ito, M.: Spectrochim. Acta *24A*, 1091 (1968).
151) Ting, C.-H.: Spectrochim. Acta *24A*, 1177 (1968).
152) Bree, A., Kydd, R. A.: J. Chem. Phys. *48*, 5319 (1968).
153) Bree, A., Kydd, R. A.: J. Chem. Phys. *51*, 989 (1969).
154) Witt, K.: Spectrochim. Acta *24A*, 1115 (1968).
155) Bree, A., Zwarich, R.: J. Chem. Phys. *51*, 912 (1969).
156) Schettino, V., Neto, N., Califano, S.: J. Chem. Phys. *44*, 2724 (1966).
157) Pecile, C., Lunelli, B.: J. Chem. Phys. *46*, 2109 (1967).
158) Brana, G. S., Adembri, G., Califano, S.: Spectrochim. Acta *22*, 1831 (1966).
159) Marzocchi, M. P., Manzelli, P., Schettino, V., Califano, S.: J. Chem. Phys. *49*, 5438 (1968).
160) Brown, C. W., Obremski, R. J., Alkins, J. R., Lippincott, E. R.: J. Chem. Phys. *51*, 1376 (1969).
161) Loisiel, J., Lorenzelli, V.: Spectrochim. Acta *23A*, 2903 (1967).
162) Jubino, R., Dellepiane, G., Zerbi, G.: J. Chem. Phys. *50*, 621 (1969).
163) Obremski, R. J., Brown, C. W., Lippincott, E. R.: J. Chem. Phys. *49*, 185 (1968).
164) Bates, J. B., Sands, D. C., Smith, W. H.: J. Chem. Phys. *51*, 105 (1969).
165) Witt, K., Mecke, R.: Ber. Bunsenges. Physik. Chem. *71*, 668 (1967).
166) Carlson, G. L., Witkowski, R. E., Fateley, W. G.: Spectrochim. Acta *22*, 1117 (1966).
167) Fukushima, K., Zwolinski, B. J.: J. Chem. Phys. *50*, 737 (1969).
168) Cadene, M.: J. Mol. Struct. *2*, 193 (1968).
169) Dilauro, C., Califano, S., Adembri, G.: J. Mol. Struct. *2*, 173 (1968).
170) Chafik, A., Mecke, R.: Z. Naturforsch. *23A*, 716 (1968).
171) Rey-Lafon, M., Trinquecoste, C., Cavagnat, R., Forel, M.-T.: J. Chim. Phys. *68*, 1532 (1971).
172) Coker, E. H., Hofer, D. E.: J. Chem. Phys. *53*, 1652 (1970).
173) Schutte, C. J. H., Ketelaar, J. A. A.: Spectrochim. Acta *17*, 1240 (1961).
174) Metselaar, R.: Thesis, University of Amsterdam, 1967.
175) Bonn, R., Metselaar, R., Van der Elsken, J.: J. Chem. Phys. *46*, 1988 (1967).
176) Kato, R., Rolfe, J.: J. Chem. Phys. *47*, 1901 (1967).
177) Manzelli, P., Taddei, G.: J. Chem. Phys. *51*, 1484 (1969).

178) Krynauw, G. N., Schutte, C. J. H.: Spectrochim. Acta 21, 1947 (1965).
179) Krynauw, G. N., Schutte, C. J. H.: Z. Phys. Chem. N.F. 55, 113 (1967).
180) Krynauw, G. N., Schutte, C. J. H.: Z. Phys. Chem. N.F. 55, 121 (1967).
181) Krynauw, G. N., Schutte, C. J. H.: Z. Phys. Chem. N.F. 55, 134 (1967).
182) McKean, D. C.: Spectrochim. Acta 23A, 2405 (1967).
183) Mann, R. H., Horrocks, W. de W.: J. Chem. Phys. 45, 1278 (1966).
184) Buckingham, A. D.: Trans Faraday Soc. 56, 753 (1960).
185) Decius, J. C.: J. Chem. Phys. 49, 1387 (1968).
186) Freedman, H., Shalom, A., Kimel, S.: J. Chem. Phys. 50, 2496 (1969).
187) Berney, C. V., Redingtom, R. L., Lin, K. C.: J. Chem. Phys. 53, 1713 (1970).
188) Grenie, Y., Lassegues, J.-C., Garrigon-Lagrange, C.: J. Chem. Phys. 53, 2890 (1970).
189) Ogden, J. S., Ricks, M. J.: J. Chem. Phys. 53, 896 (1970).
190) Tursi, A. J., Nixon, E. R.: J. Chem. Phys. 53, 518 (1970).
191) Pacansky, J., Calder, V.: J. Chem. Phys. 53, 4519 (1970).
192) Allavena, M., Rysnik, R., White, D., Calder, V., Mann, D. E.: J. Chem. Phys. 50, 3399 (1969).
193) Guillory, W. A., Hunter, C. E.: J. Chem. Phys. 50, 3516 (1969).
194) King, S. T.: J. Chem. Phys. 49, 1321 (1968).
195) Milligan, D. E., Jacox, M. E.: J. Chem. Phys. 53, 2034 (1970). – See the neutron results of Stirling, G. E., Waddington, T. C.: J. Chem. Phys. 52, 2730 (1970).
196) Acquista, N., Schoen, L. J.: J. Chem. Phys. 53, 1290 (1970).
197) Keyser, L. F., Robinson, G. W.: J. Chem. Phys. 45, 1694 (1966).
198) Von Holle, W., Robinson, D. W.: J. Chem. Phys. 53, 3768 (1970).
199) Davies, M. (ed.): Infra-red spectroscopy and molecular structure. Amsterdam: Elsevier Publishing Co. 1962.
200) Brooker, M. H.: J. Chem. Phys. 53, 4100 (1970).
201) Devlin, J. P., Pollard, G., Frech, R.: J. Chem. Phys. 53, 4147 (1970).
202) Ahlijah, G. E. B. Y., Mooney, E. F.: Spectrochim. Acta 25A, 619 (1969).
203) Deane, A. M., Richards, E. W. T., Stephen, I. G.: Spectrochim. Acta 22, 1253 (1966).
204) Ferraro, J. R., Postmus, C., Mitra, S. S.: Phys. Rev. 174, 983 (1968).
205) Ferraro, J. R., Postmus, C., Mitra, S. S., Hoskins, C. J.: Appl. Opt. 9, 5 (1970).
206) Ferraro, J. R., Mitra, S. S., Postmus, C., Hoskins, C. J., Siwiec, E. C.: Appl. Spectry. 24, 187 (1970).
207) Ferraro, J. R.: J. Chem. Phys. 53, 117 (1970).
208) Blinc, R., Ferraro, J. R., Postmus, C.: J. Chem. Phys. 51, 732 (1969).
209) Asell, J. F., Nicol, M.: J. Chem. Phys. 49, 5395 (1968).
210) Fong, M. Y., Nichol, M.: J. Chem. Phys. 49, 5395 (1968).
211) Jamieson, J. C.: J. Geol. 65, 334 (1957).
212) Davies, B. L.: Science 145, 489 (1964).
213) Rapoport, E.: J. Phys. Chem. Solids 27, 1349 (1966).
214) Heyns, A. M., Schutte, C. J. H., Scheuermann, W.: J. Mol. Struct. 9, 271 (1971).
215) Ketelaar, J. A. A., Schutte, C. J. H.: Spectrochim. Acta 17, 815 (1961).
216) Schutte, C. J. H.: Spectrochim. Acta 16, 1054 (1960).
217) Subratová, V.: Z. Anorg. Allgem. Chem. 350, 211 (1967).
218) Willett, R. D., Dwiggins, C., Jr., Kruh, R. F., Rundle, R. E.: J. Chem. Phys. 38, 2429 (1963).
219) Heyns, A. M., Schutte, C. J. H.: J. Mol. Struct. 8, 339 (1971).
220) Schutte, C. J. H., Van Rensburg, D. J. J.: J. Mol. Struct. 10, 481 (1971).

221) Caron, A. P., Huettner, D. J., Ragle, J. L., Sherk, L., Stengle, R. T.: J. Chem. Phys. *47*, 2577 (1967). −*Erratum*, J. Chem. Phys. *48*, 4331 (1968).
222) Schutte, C. J. H., Heyns, A. M.: J. Mol. Struct. *5*, 37 (1970).
223) Castellucci, E., Sbrana, G., Verderame, F. D.: J. Chem. Phys. *51*, 3762 (1969).
224) May, I., Pace, E. L.: J. Mol. Struct. *25A*, 1903 (1969).
225) Brüesch, P., Günthard, Hs. H.: Spectrochim. Acta *22*, 877 (1966).
226) Craft, W. L., Slutsky, L. J.: J. Chem. Phys. *49*, 638 (1968).
227) Sato, Y.: J. Chem. Phys. *53*, 887 (1970).
228) Wiener, E. (Avnear), Levin, S., Pelah, I.: J. Chem. Phys. *52*, 2881 (1970).
229) Wiener, E. (Avnear), Levin, S., Pelah, I.: J. Chem. Phys. *52*, 2891 (1970).
230) Swanson, B. I., Jones, L. H.: J. Chem. Phys. *53*, 3761 (1970). − See also J. Chem. Phys. *55*, 4174 (1971) for a normal coordinate analysis.
231) Jones, L. H.: J. Chem. Phys. *36*, 1209 (1962).
232) Jones, L. H.: J. Chem. Phys. *41*, 856 (1964).
233) Bloor, D.: J. Chem. Phys. *41*, 2573 (1964).
234) McAllister, W. A.: J. Chem. Phys. *52*, 2786 (1970).
235) Loehr, F. M., Long, T. V.: J. Chem. Phys. *53*, 4182 (1970).
236) Jones, L. H.: J. Chem. Phys. *22*, 965 (1955).
237) Jones, L. H.: J. Chem. Phys. *25*, 379 (1956).
238) Jones, L. H.: J. Chem. Phys. *26*, 1578 (1957).
239) Bottger, G. L.: Spectrochim. Acta *24A*, 1821 (1968).
240) Chadwick, B. M., Frankiss, S. G.: J. Mol. Struct. *2*, 281 (1968).
241) Griffith, W. P., Turner, G. T.: J. Chem. Soc. *1970A*, 858.
242) Krasser, W.: Ber. Bunsenges. Physik. Chem. *74*, 476 (1970).
243) Hartman, K. O., Miller, F. A.: Spectrochim. Acta *24A*, 669 (1968).
244) Hoard, J. L., Nordsieck, H. H.: J. Am. Chem. Soc. *61*, 2853 (1939).
245) Hoard, J. L., Martin, W. J., Smith, M. E., Whitney, J. F.: J. Am. Chem. Soc. *76*, 3820 (1954).
246) Stammreich, H., Scala, O.: Z. Elektrochem. *64*, 741 (1960).
247) Stammreich, H., Scala, O.: Z. Elektrochem. *65*, 149 (1961).
248) Siebert, H., Eysel, H. H.: J. Mol. Struct. *4*, 29 (1969).
249) Nakagawa, I., Shimanouchi, T.: Spectrochim. Acta *22*, 1707 (1966).
250) Nakagawa, I., Shimanouchi, T.: Spectrochim. Acta *23A*, 2099 (1967).
251) Debeau, M., Poulet, H.: Spectrochim. Acta *25A*, 1553 (1969).
252) Hartley, D., Ware, M. J.: J. Chem. Soc. *1969A*, 138.
253) Long, T. V., Huege, F. R.: Chem. Commun. *1968*, 1239.
254) Bunker, P. R.: Mol. Phys. *9*, 247 (1965).
255) Longuet-Higgins, H. C.: Mol. Phys. *6*, 445 (1963).
256) Murrell, J. N.: J. Chem. Soc. *1969A*, 297.
257) Lane, A. P., Sharp, D. W. A.: J. Chem. Soc. *1969A*, 2942.
258) Adams, D. M., Newton, D. C.: J. Chem. Soc. *1969A*, 2998.
259) Beattie, I. R., Livingston, K. M. S., Ozin, G. A., Reynolds, D. J.: J. Chem. Soc. *1969A*, 958.
260) Buttery, H. J., Keeling, G., Kettle, S. F. A., Paul, I., Stamper, P. J.: J. Chem. Soc. *1969A*, 2077.
261) Buttery, H. J., Keeling, G., Kettle, S. F. A., Paul, I., Stamper, P. J.: J. Chem. Soc. *1969A*, 2225.
262) Kittelberger, J. S., Hornig, D. F.: J. Chem. Phys. *46*, 3099 (1967).
263) Wang, C. H., Fleury, P. A.: J. Chem. Phys. *53*, 2243 (1970).
264) Friedrich, T. B., Carlson, R. C.: J. Chem. Phys. *53*, 4441 (1970).
265) Savoie, R., Anderson, A.: J. Chem. Phys. *44*, 548 (1966).

C. J. H. Schutte

266) Ito, M., Suzuki, M., Yokoyama, T.: J. Chem. Phys. *50*, 2949 (1969).
267) De Bettignies, B., Wallert, F.: Compt. Rend. *271B*, 640 (1970).
268) Herzog, T., Schwab, G.-M.: Z. Phys. Chem. N.F. *66*, 190 (1969).
269) Johansen, H.: Z. Naturforsch. *21A*, 836 (1966).
270) Ewing, G. E., Pimentel, G. C.: J. Chem. Phys. *35*, 925 (1961).

Received February 7, 1972

Laser Raman Spectroscopy of the Solid State

Prof. Dr. Josef Brandmüller und Dr. Heinz W. Schrötter

Sektion Physik der Universität München

Contents

I. The Different Causes of the Raman Effect

In the elementary Raman scattering process light is inelastically scattered in a material system. An incident photon of frequency ω_i is annihilated and a scattered photon of requency ω_s is created instead. In addition a quantum of frequency ω is created in the Stokes case or annihilated in the anti-Stokes case.

Already in 1923 Smekal [1] postulated the existence of frequency-shifted scattered radiation on the basis of quantum theory. He gave the energy relation

$$\hbar\omega_i + E_k = E_n + \hbar\omega_s, \tag{I.1}$$

where E_k is the energy of some state of the material system before the scattering process and E_n that of another state after the process, and $\hbar = h/2\pi$ with h for Planck's constant. Accordingly the energy of the material system can be changed by the scattering process. Smekal considered only the different energy levels of a molecule, for instance vibrational and rotational levels. Today the meaning of the term "energy level" must be understood much more generally.

The low quantized excitation levels or elementary excitations of the material system are also called quasi-particles in solid state physics by analogy with the elementary particles in quantum-field theory [2,3].

Up to the present time the following elementary excitations are known to be of importance for the Raman effect: rotation of a whole molecule, vibrations of the atoms in a molecule, and coupled states of vibration and rotation. An electronic level can also lead to frequency-shifted scattered radiation. Rasetti [4] observed the electronic Raman effect in nitric oxide as early as 1930. It was not until 1963 that interest in this almost forgotten effect was revived by its investigation in a single crystal of $PrCl_3$ [5]. Since then it has become important in solid-state physics for the determination of electronic energy levels in single crystals, mainly through the work of Koningstein and his group [6]. The rotational structure of the electronic Raman band of nitric oxide could also be resolved [7].

The elementary excitations mentioned so far are not related in any special way to the solid state and will therefore not be treated in this article. We will discuss here the following low-lying quantized excitations or quasi-particles which have been investigated by Raman spectroscopic methods: phonons, polaritons, plasmons and coupled plasmon-phonon states, plasmaritons, magnons, and Landau levels. Finally, phase transitions were also studied by light scattering experiments; however, they cannot be dealt with in this article.

When the power of the exciting radiation is raised into the megawatt range, nonlinear Raman effects are observed, namely the stimulated Raman effect, the inverse Raman effect (Stoicheff absorption), and the hyper-Raman effect. The results of such experiments with single crystals will be discussed in the last chapter, with special emphasis on stimulated Raman scattering from polaritons.

II. Linear Raman Effect

A. Long-Wave Optical Phonons and Polaritons

1. Longitudinal Optical and Acoustical Phonons

A simple example which can be used to explain the basic principles is the linear infinite chain with two atoms in the unit cell.

a) The Linear AB Chain

We assume that a longitudinal wave propagates in the AB chain (Fig. 1). By longitudinal, we mean that the elongation u of the atoms or ions coincides with the direction of wave propagation, *e.g.* the x-axis.

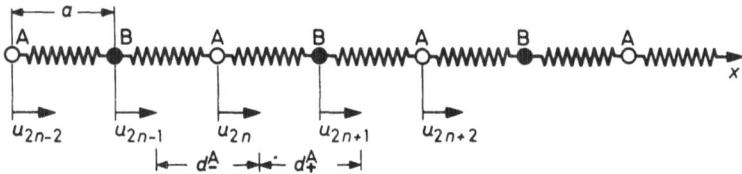

Fig. 1. Linear AB chain

All A atoms are labelled with the even numbers $2n$, all B atoms with the odd numbers $2n \pm 1, n = 1,2,3 \ldots$ When the distances of the $2n^{\text{th}}$ atom from both its neighbours are equal at any particular moment, no force is acting on this atom taking only nearest-neighbour interactions into account, which is a good approximation. The force on the $2n^{\text{th}}$ A atom depends on the difference of the distances d^A of this atom from its two nearest neighbours of type B, namely $d_+^A - d_-^A$. The indices denote the positive and negative x-directions, respectively.

The distance A—B in the positive x-direction is

$$d_+^A = u_{2n+1} - u_{2n} + a$$

and in negative x-direction

$$d_-^A = u_{2n} - u_{2n-1} + a.$$

The equation of motion for the atom A_{2n} is therefore

$$m_A \ddot{u}_{2n} = f(d_+^A - d_-^A) = f(u_{2n+1} + u_{2n-1} - 2u_{2n}), \qquad (\text{II.1})$$

and hence for the atom B_{2n+1}

$$m_B \ddot{u}_{2n+1} = f(d_+^B - d_-^B) = f(u_{2n+2} + u_{2n} - 2u_{2n+1}). \qquad (\text{II.2})$$

89

These equations of motion are solved with a harmonic ansatz. $u_{2n}(x)$ and $u_{2n+1}(x)$ are obtained as a function of x at $x_A = 2na$ and $x_B = (2n \pm 1) \cdot a$ because of the discontinuity of the chain.

For a harmonic plane wave propagating in the x-direction one obtains

$$u_{2n} = U_A e^{-i(\omega t - kx_A)} = U_A e^{-i(\omega t - 2nka)}$$

$$u_{2n\pm 1} = U_B e^{-i(\omega t - kx_B)} = U_B e^{-i[\omega t - (2n\pm 1)ka]} \tag{II.3}$$

where U_A and U_B are the amplitudes, ω the angular frequency, and $k = 2\pi/\lambda$ the magnitude of the wave vector, with λ the wavelength.

In order to find a dispersion relation $\omega = \omega(k)$, a system of equations for the unknown amplitudes U_A and U_B is derived. The secular equation leads immediately to the dispersion relation (see [8,9])

$$\omega^2 = f\left(\frac{1}{m_A} + \frac{1}{m_B} \pm \sqrt{\left(\frac{1}{m_A} + \frac{1}{m_B} \right)^2 - \frac{4 \sin^2 ka}{m_A m_B}} \right), \tag{II.4}$$

From this equation several conclusions can be drawn.

b) The Brillouin Zones

Because it contains $\sin^2 ka$, the function $\omega = \omega(k)$ is periodic. All possible eigenfrequencies are obtained when in Eq. (II.4) $\sin^2 ka$ is varied only in the range $0 \leqslant ka \leqslant \pi/2$. The magnitude of the wave vector is thus

$$0 \leqslant k \leqslant \pi/2a.$$

This range in k space is called the first Brillouin zone (BZ 1), see Fig. 2; higher values of k give further Brillouin zones with identical values for $\omega = \omega(k)$. The

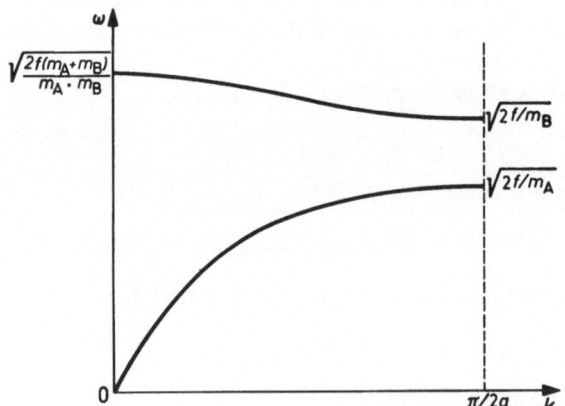

Fig. 2. Dispersion curves of the optical and acoustical branch in BZ 1 for the linear AB chain

magnitude of the wave vector k cannot become negative. $k = 0$ denotes the center of BZ1. The other half of BZ1 is obtained when the wave vector $\vec{k}$ points in the negative x-direction.

c) Dispersion in the Center of BZ 1

Near the center of BZ 1 $k \ll \pi/2a$. When the root in (II.4) is expanded and nonlinear terms are neglected, one obtains

$$\omega^2\left(k \ll \frac{\pi}{2a}\right) \approx \frac{f}{m_A m_B}\left[m_A + m_B \pm (m_A + m_B) \mp \frac{2m_A m_B}{m_A + m_B} k^2 a^2\right]. \qquad (II.5)$$

The different signs lead to the two cases:

a) Upper signs. The term with $k^2 a^2$ can be neglected and we are left with

$$\omega_+^2\left(k \ll \frac{\pi}{2a}\right) \approx \frac{2f}{m_A m_B}(m_A + m_B). \qquad (II.6)$$

In this approximation ω becomes independent of k, therefore this branch of the dispersion curve is horizontal in the center of the BZ 1. This is the *optical branch*.

b) Lower signs. In this case

$$\omega_-^2\left(k \ll \frac{\pi}{2a}\right) \approx \frac{2f}{m_A + m_B} k^2 a^2. \qquad (II.7)$$

ω_- is linearly dependent on k and goes to zero with k. This solution corresponds to the *acoustical branch*.

d) Eigenfrequencies at the Edge of BZ 1

For $k = \pi/2a$ Eq. (II.4) becomes

$$\omega^2\left(k = \frac{\pi}{2a}\right) = \frac{f}{m_A m_B}[m_A + m_B \pm (m_A - m_B)]. \qquad (II.8)$$

Again two cases are to be distinguished.

a) Upper sign.

$$\omega_+^2\left(k = \frac{\pi}{2a}\right) = \frac{2f}{m_B}. \qquad (II.9)$$

The frequency of the optical branch at the edge of BZ 1

$$\omega_+\left(k = \frac{\pi}{2a}\right) = \sqrt{2f/m_B}$$

is therefore smaller than

$$\omega_+(k = 0) = \sqrt{2f\left(\frac{1}{m_A} + \frac{1}{m_B}\right)},$$

the frequency of the same branch in the center of BZ 1.

b) Lower sign.

$$\omega_-^2\left(k = \frac{\pi}{2a}\right) = \frac{2f}{m_A}. \tag{II.10}$$

At this point a decision must be made as to whether atom A or atom B has the greater mass. If ω_+ is to denote the optical branch throughout BZ 1, atom A must have the greater mass: $m_A > m_B$.

As regards the notation "acoustical" and "optical" for phonons, what is important is the ratio between the vibrational amplitudes in the acoustical and optical branches. The amplitude ratio U_A/U_B for the different values of ω and k can be calculated from Eqs. (II.1) to (II.3). We thus justify the use of the terms acoustical and optical branches as follows.

e) Acoustical Branch

According to Eq. (II.7), $\omega_- = 0$ for $k = 0$ in the center of BZ 1. With these values Eqs. (II.1)−(II.3) lead to the relation $U_A = U_B$. This means that both sets of atoms vibrate with the same amplitude and in phase (because they have the same sign). A translation of the whole chain results which corresponds to an acoustical wave with $\lambda = \infty$. This is called a longitudinal acoustical branch (LA).

At the edge of BZ 1, according to Eq. (II.10), the eigenfrequency $\omega_- = \sqrt{2f/m_A}$ for $k = \pi/2a$. Insertion of these values in Eqs. (II.3) and (II.2) gives $(2fm_B/m_A - 2f)U_B = 0$. For $m_A \neq m_B$ it follows that $U_B = 0$ and U_A is arbitrary. This means that atoms with the smaller mass m_B remain in the equilibrium position and only the heavier A atoms vibrate. For $k = \pi/2a$ it follows from $k = 2\pi/\lambda$ that the wavelength $\lambda = 4a$. When atoms A and B are charged, no change of electric dipole moment is associated with this vibration, therefore it is also called acoustical.

f) Optical Branch

In the center of BZ 1, *i.e.* for $k = 0$, ω_+ is given by Eq. (II.6). By insertion in Eqs. (II.1)−(II.3) one obtains $U_A/U_B = -m_B/m_A$.

The amplitude is inversely proportional to the mass and the A and B atoms vibrate with opposite phases.

At the edge of BZ 1 $(k = \pi/2a)$ $\omega_+ = \sqrt{\frac{2f}{m_B}}$ according to Eq. (II.9). From Eqs. (II.1) and (II.3) it therefore follows that $(m_A/m_B - 1)U_A = 0$. For

$m_A \neq m_B$, therefore, $U_A = 0$ and U_B is arbitrary; this means that the heavier A atoms remain in equilibrium. For the ω_+ modes the center of gravity of the A and B atoms does not move throughout the whole of BZ 1. When the A and B atoms have different electric charges, a change of electric dipole moment is associated with these vibrations except at the edge of BZ 1. Therefore this branch is called the longitudinal optical branch (LO). This notation is also valid, when the atoms are not charged, provided they vibrate in the manner described above. All exclusively Raman-active and also all "silent" modes (*i.e.* IR and Raman-inactive) are called "optical" modes.

2. Transverse Optical and Acoustical Phonons

So far only one degree of freedom of the vibration has been considered, namely, in the direction of the wave vector. The removal of this restriction gives transverse optical and acoustical phonons. For these, the atoms or ions move perpendicular to the direction of wave propagation. Again, there are two possibilities. When A and B atoms vibrate in phase, there is no change of dipole moment and one speaks of a transverse acoustical phonon (TA). However, for a vibration with opposite phases in the A and B atoms, the electric dipole moment changes so that we have a transverse optical phonon (TO).

3. Space Lattice

The atoms or ions of a lattice are arranged in the three directions of space. The regularity of the array of the atoms in ideal crystals permits the subdivision of the crystal space into equal and equally oriented regions of space, the so-called elementary cells. Each cell contains the same complex of atoms with the same orientation in space. Such an elementary cell may contain s atoms. Each atom has three degrees of freedom, one in each of the directions of the three coordinate axes. Therefore a space lattice has $3\,s$ eigenfrequencies or modes. Of these $3\,s$ modes for $k = 0$ (*i.e.* in the center of BZ 1) three correspond to the translations into the directions of the coordinate axes. These have the frequency $\omega = 0$, which corresponds to the frequencies of the acoustical branches according to Eq. (II.7) for $k = 0$. The $3\,s$ eigenfrequencies of a crystal with s atoms in the elementary cell correspond to $3\,s - 3$ optical and 3 acoustical branches.

A strict classification into longitudinal and transverse vibrations is possible only in special cases, *e.g.* for the linear chain because of its high symmetry. In a space lattice, on the other hand, only the phonons propagating in the directions of the principal axes of the crystal are purely longitudinal or transverse, whereas phonons with an oblique k vector generally have a mixed character. This fact is decisive for the choice of useful scattering geometries for the assignment of the phonons to certain symmetry species.

4. Observation of Phonons by Light Scattering

Raman scattering can be considered as inelastic scattering of a photon by the phonons of a crystal. For a one-phonon process, energy conservation leads to the relations $\omega_i = \omega_s + \omega$ in the Stokes case and $\omega_i = \omega_s - \omega$ in the anti-Stokes case. The incident and the scattered photons, denoted by i and s, respectively, also have momentum. The momentum I of a photon is calculated from $I = m \cdot c$ and $\hbar\omega = mc^2$ as $I = \hbar\omega/c = \hbar k$. Because the momentum is a vector, one must write $\vec{I} = \hbar\vec{k}$. Now there arises the question whether a phonon also has momentum. With a mechanical wave in a crystal there is no motion of the center of gravity, therefore a phonon has no momentum in the usual sense. In spite of this a quasi-momentum can be ascribed to it. This question has been treated by Süssmann [10]. One may write down a relation for the k vectors which corresponds to conservation of momentum if all wave vectors of the phonons are within BZ 1, namely $\vec{k}_i = \vec{k}_s + \vec{k}$ for the Stokes case and $\vec{k}_i = \vec{k}_s - \vec{k}$ for the anti-Stokes case.

For the following argument the order of magnitude of the wave vectors of the phonons created or annihilated by light scattering is of special importance. Usually visible light is used as the exciting radiation for the Raman effect. Let us assume $\lambda_i \approx 500$ nm $\stackrel{\wedge}{=} 5 \cdot 10^{-5}$ cm. Then the absolute magnitude of the wave vector of the incident photon $k_i = 2\pi/\lambda_i \approx 10^5$ cm^{-1}. The wave number of the phonons observed in the Raman effect is in the order of 10 to 1000 cm^{-1}, so that the wave numbers and the magnitudes of the wave vectors of the incident and scattered photons differ only slightly. For 90 ° scattering, which is used in most cases, the wave vector triangle is nearly isosceles.

The magnitude of the phonon wave vector is then $k \approx \sqrt{2}\, k_i$, i.e. k is also approximately 10^5 cm^{-1}. At the edge of BZ 1 the phonon wave vector has the magnitude $k = \pi/2a$. The lattice constant of a crystal is approximately $a \approx 0.1$ nm, so that we have $k \approx 10^8$ cm^{-1} at the edge of BZ 1. Therefore in light scattering experiments the phonon wave vector is about three orders of magnitude smaller than at the edge of BZ 1, i.e. it is near its center.

The scattering at the acoustical branches leads to Brillouin scattering which will not be treated here. According to elementary theory used until now, the optical branches have no dispersion in the region of small k values. Therefore one might assume that the results of group theory for $k = 0$ might be fully applicable for finite but small values of k, as observed in light-scattering experiments. However, this is not true, as the following considerations will show.

It should be emphasized, however, that we are dealing in this article mainly with one-phonon processes, i.e. fundamentals. For multiple phonon processes, i.e. overtones and combinations which are observed in second-order Raman scattering, the whole of BZ 1 comes into play. Because the sum or difference of two large $\vec{k}$ vectors can have a relatively small magnitude, it can fulfill the condition that the magnitude of the resulting phonon wave vector is of the same order of magnitude as k_i.

The following considerations are only valid for fundamentals.

5. The Dispersion of Long-Wave Polar Optical Phonons in Diatomic Cubic Crystals

This was first investigated by Huang [11]. Long-wave optical phonons are those with small k values. A polar phonon is an infrared-active phonon. Polar phonons therefore can only be observed in the Raman effect for crystals having no center of symmetry in the elementary cell. For centro-symmetric crystals the rule of mutual exclusion applies: infrared-active phonons are forbidden in Raman scattering and vice versa. The elementary cells of NaCl and LiF have a center of symmetry, but GaP has none. The following considerations may therefore be applied to GaP as an example. This crystal has two atoms in the elementary cell and is cubic. It can be treated as an optically isotropic medium.

a) Huang's Equations

For the polar vibrations of the two sublattices against each other, according to Huang the following equation of motion is valid

$$\ddot{\vec{w}} = b_{11}\vec{w} + b_{12}\vec{E} . \tag{II.11}$$

Without the second term on the right-hand side this is the equation of a harmonic oscillator. $\vec{w}$ is the (mass-loaded) vector of the relative elongation of the two sublattices. For dimensional reasons b_{11} must be the square of a frequency. From the solution of the harmonic equation one obtains

$$b_{11} = -\omega_T^2, \tag{II.12}$$

where ω_T designates for the moment the eigenfrequency of the system observable in infrared absorption. The second term in Eq. (II.11) constitutes an additional repelling force originating from an electrical field $\vec{E}$. This is not an external field, but one that is created by the vibration of the two sublattices with opposite charges. Thus a macroscopic electrical field is formed in the crystal with a periodic time dependence ($e^{i\omega t}$, with ω the frequency of the phonon). Its frequency is, however, very small compared to that of a light wave. Therefore the electric field originating from the mechanical vibration may be regarded as quasi electrostatic. It is a long-range electric field extending over many elementary cells. This $\vec{E}$ field reacts on the motion of the ions, so that the frequency of the LO phonon will be changed under its influence. This is the background for the Lyddane-Sachs-Teller relation to be treated below. For transverse optical vibrations the origin of an $\vec{E}$ field is less obvious, but it is also present and its reaction on the eigenfrequency of the TO phonon later gives rise to the polaritons.

In addition to Eq. (II.11) Huang writes a further one:

$$\vec{P} = b_{21}\vec{w} + b_{22}\vec{E} .$$ (II.13)

$\vec{P}$ is the macroscopic polarization. It consists of a lattice polarization $b_{21}\vec{w}$ originating from the electric dipole moment arising from the mutual displacement of the two sublattices, and of a second term $b_{22}\vec{E}$ originating from the pure electron polarization. According to definition, $\vec{P}$ and $\vec{E}$ are connected by

$$\vec{P} = \frac{\epsilon - 1}{4\pi} \vec{E} ,$$ (II.14)

where ϵ is the dielectric constant. The lattice vibration is assumed to be harmonic:

$$\vec{w} = \vec{w}_0 \cdot e^{-i(\omega t - \vec{k} \cdot \vec{r})} .$$ (II.15)

From energy conservation [12] it follows that $b_{12} = b_{21}$. From Eqs. (II.11) to (II.15) the following macroscopic expressions can be derived for the coefficients $b_{\mu\nu}$:

$$b_{12} = \sqrt{\frac{\epsilon_0 - \epsilon_\infty}{4\pi}}\omega_T; \quad b_{22} = \frac{\epsilon_\infty - 1}{4\pi} .$$ (II. 16)

ϵ_0 is the static dielectric constant; ϵ_∞ the dielectric constant for frequencies much greater than ω_T. $\epsilon_\infty = n^2$ where n is the optical refractive index. Generally ϵ is a function of ω which can be written for a single absorption frequency ω_T as

$$\epsilon(\omega) = \frac{\epsilon_0 \omega_T^2 - \epsilon_\infty \omega^2}{\omega_T^2 - \omega^2} \equiv \frac{\omega_L^2 - \omega^2}{\omega_T^2 - \omega^2} \epsilon_\infty.$$ (II.17)

ω_L is the frequency where the function $\epsilon(\omega)$ is zero. Later it will acquire another significance.

b) Retardation Effect

The relations derived up to this point are not sufficient to solve the problem of the dispersion of the optical branch for small k values. The retardation effect must also be taken into account. Because of the finite velocity of electromagnetic waves, the forces at a certain point of time and space in a crystal are determined by the states of the whole crystal at earlier times. A precise description of the dispersion effect therefore requires the introduction of Maxwell's equations. With a harmonic ansatz for $\vec{E}$ and $\vec{P}$ which is analogous to Eq. (II.15) they lead to the relation

$$\vec{E} = \frac{4\pi}{k^2 - \frac{\omega^2}{c^2}} \left[\frac{\omega^2}{c^2} \vec{P} - (\vec{k} \cdot \vec{P}) \vec{k} \right].$$ (II.18)

From Eqs. (II.11), (II.13), (II.14), and (II.18) finally follows

$$\epsilon(\vec{k} \cdot \vec{E}) = 0 \ . \tag{II.19}$$

This relation can be fulfilled in two ways.

c) Dispersion of the LO mode

$\epsilon = 0$ leads according to Eq. (II.17) to the relation

$$\omega = \omega_L = \sqrt{\frac{\epsilon_0}{\epsilon_\infty}} \, \omega_T . \tag{II.20}$$

One can show, that in this case $\vec{k}$ is parallel to $\vec{E}$, $\vec{P}$, and also to $\vec{w}$, i.e. this solution is the dispersion relation of the LO mode. It is remarkable that ω_L, which now turns out to be the frequency of the LO mode, does not depend on k. ω_L has no dispersion in the center of BZ 1 and is always greater than ω_T because $\epsilon_0 > \epsilon_\infty$. Eq. (II.20) is called the Lyddane-Sachs-Teller relation.

d) Dispersion of the TO mode

The other possible way to fulfill Eq. (II.19) is $\vec{k} \cdot \vec{E} = 0$. One can show in this case that $\vec{k}$ is perpendicular to $\vec{E}$ and $\vec{w}$, therefore this is the solution for the TO mode. After a short calculation from $\vec{k} \cdot \vec{E} = 0$ we obtain the dispersion relation

$$\frac{k^2 c^2}{\omega^2} = \epsilon(\omega) \, , \tag{II.21}$$

Because ϵ is a function of ω according to Eq. (II.17), (II.21) gives implicitly the function $\omega = \omega(k)$. An explicit calculation of this dispersion relation is possible, but leads to somewhat awkward expressions. The result is shown graphically in Fig. 3.

The full horizontal line is the dispersion line of the LO mode, Eq. (II.20). From Eq. (II.21) two branches result. The lower branch starts at $\omega = 0$ with the slope $c/\sqrt{\epsilon_0}$. For very small k values the curve closely follows a straight line with this slope. This dashed straight line represents the dispersion relation of photons in a dielectric with the refractive index $\sqrt{\epsilon_0}$. The dispersion branch then departs from this straight line for greater k values and asymptotically approaches ω_T, the infrared absorption frequency. Therefore the lower dispersion curve is photon-like for small k values and phonon-like for large ones. In the transition region it represents a mixed state between an electromagnetic and a mechanical wave. These mixed photon-phonon states are called "polaritons" [13].

The upper dispersion branch begins at ω_L for $k = 0$ and is at first phonon-like, then increasingly assumes photon character. For large k values it is asymptotic to a straight line with the slope $c/\sqrt{\epsilon_\infty}$ which describes the dispersion

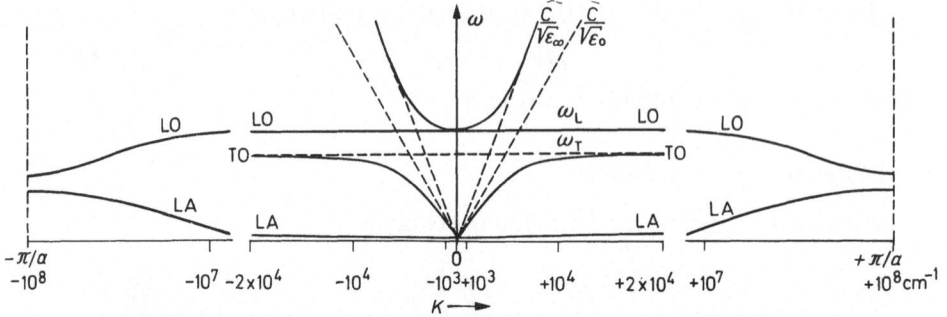

Fig. 3. Dispersion curves of the long-wavelength optical phonons, photons, and polaritons in the centre of BZ 1. In order to demonstrate the connection with the dispersion effects in the region $10^7 < k < 10^8$ cm^{-1}, the branches of an LO and an LA phonon in this region have been added in a different linear scale. The figures correspond to a cubic lattice with two atoms in the unit cell [35]

of a photon in an optical medium with the refractive index $n = \sqrt{\epsilon_\infty}$. For intermediate k values we again have a mixed state, namely a polariton.

6. Dispersion of Biaxial, Polyatomic Crystals

Merten [14] has extended the theory for the general case of a biaxial, polyatomic crystal. His theory is worked out in detail for crystals of the orthorhombic system and all crystal classes of higher symmetry, but not yet for crystals of the monoclinic and triclinic systems.

a) The General Dispersion Formula

The general dispersion formula obtained for the coupling of the vibrational equations with the Maxwell field can be brought into the form of Fresnel's wellknown equation for the wave normal from crystal optics. It is usually written in the form

$$\sum_{\eta=1}^{3} s_\eta^2 / \left(\frac{1}{n^2} - \frac{1}{\epsilon_\eta} \right) = 0 . \tag{II.22}$$

When used as the dispersion formula for the phonons and polaritons in orthorhombic crystals, the symbols in Eq. (II.22) have the following meaning: $\eta = 1, 2, 3$ designates the three directions of the principal orthogonal axes. s_η are the direction cosines of the normalized wave vector $\vec{s} = \vec{k}/k$ with respect to the three principal axes of the crystal. If the unit vectors in the directions of these three principal axes are designated $\vec{e}_1, \vec{e}_2, \vec{e}_3$, one can write

$$\vec{s} = s_1 \vec{e}_1 + s_2 \vec{e}_2 + s_3 \vec{e}_3 = \sum_{\eta=1}^{3} s_\eta \vec{e}_\eta \ . \qquad \text{(II.23)}$$

The refractive index n is not regarded as a constant as in Fresnel's equation, but as an abbreviation for $k \cdot c/\omega$. This can be understood as follows: in an optical medium the phase velocity of light is c/n, where c is the velocity of light in the vacuum. Now the phase velocity is connected with wavelength and frequency through the relation $c/n = \lambda \cdot \nu$. On the other hand $k = 2\pi/\lambda$. This leads to $c/n = \omega/k$ and finally

$$n = \frac{k \cdot c}{\omega} \ . \qquad \text{(II.24)}$$

ϵ_η designates the frequency-dependent dielectric constants in the three principal directions of the crystal. As a generalization of the expression for the frequency-dependent dielectric constant of a cubic crystal with two atoms in the elementary cell given in Eq. (II.17), following Kurosawa [15] the ansatz is made for an orthorhombic crystal with any number of atoms in the elementary cell

$$\epsilon_\eta = \epsilon_{\infty\eta} \prod_{i=1}^{m_\eta} \frac{\omega_{Li}^2 - \omega^2}{\omega_{Ti}^2 - \omega^2} \ . \qquad \text{(II.25)}$$

$\epsilon_{\infty\eta}$ are the dielectric constants at high frequencies, *i.e.* the squares of the refractive indices in the respective directions η. $i = 1, \ldots, m_\eta$ is the index running over the infrared-active transverse absorption frequencies ω_{Ti} measured in direction η. The ω_{Li} mathematically are the frequencies where the function $\epsilon_\eta = \epsilon_\eta(\omega)$ becomes zero.

The physical significance of the ω_{Li} is that of the LO frequencies belonging to the ω_{Ti}. After insertion of Eqs (II.24) and (II.25) into Eq. (II.22) the latter represents an implicit dispersion relation $\omega = \omega(k)$. The form of Eq. (II.22) is simple but has the disadvantage that the solution $\epsilon_\eta = n^2$ must be rejected because of the vanishing denominator. It is better to remove the denominator. A short calculation leads to the less simple expression

$$\sum_{\eta=1}^{3} \left\{ \epsilon_\eta s_\eta^2 \prod_{\substack{\zeta=1 \\ \zeta \neq \eta}}^{3} (\epsilon_\zeta - n^2) \right\} = 0. \qquad \text{(II.22a)}$$

This relation is of central importance for the dispersion of phonons and polaritons. Special cases can immediately be derived from it. The case of the cubic diatomic crystal treated above follows from $\epsilon_1 = \epsilon_2 = \epsilon_3 = \epsilon$. Because $\vec{s}$ is a unit vector, $s_1^2 + s_2^2 + s_3^2 = 1$. Therefore from Eq. (II.22a) follows

$$\epsilon (\epsilon - n^2)^2 = 0. \qquad \text{(II.19a)}$$

From Eq. (II.17) follows for $\epsilon = 0$ Eq. (II.20). From the other possible way to fulfill Eq. (II.19a) there follows with Eq. (II.24) the condition (II.21) already obtained above.

b) Cubic, Polyatomic Crystals

Compared with the case of the diatomic crystal only the expression for $\epsilon(\omega)$ is changed. From Eq. (II.25) we obtain for this case

$$\epsilon(\omega) = \epsilon_\infty \prod_{i=1}^{m} \frac{\omega_{Li}^2 - \omega^2}{\omega_{Ti}^2 - \omega^2} . \tag{II.26}$$

m is the number of infrared active phonons. Barker [16] showed that from the condition $\epsilon = 0$ there follows a generalized Lyddane-Sachs-Teller relation

$$\prod_{i=1}^{m} \omega_{Li}^2 = \frac{\epsilon_0}{\epsilon_\infty} \prod_{i=1}^{m} \omega_{Ti}^2. \tag{II.27}$$

Consequently, regarding only Eq. (II.27) nothing can be said about the single LO frequencies, but only about their product. The independence of k is retained. The behaviour of the polaritons follows from Eq. (II.21) where Eq.

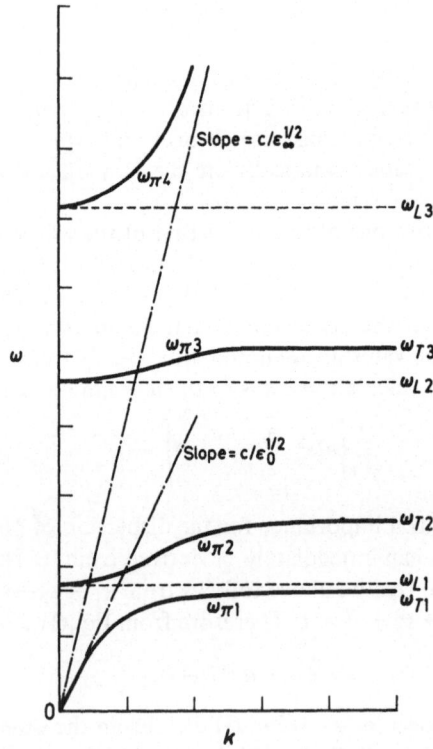

Fig. 4. Polariton dispersion curves for a cubic crystal with three infrared active phonons [17]

(II.26) is to be inserted for $\epsilon(\omega)$. The form of the dispersion curves $\omega = \omega(k)$ described implicitly by this equation is reproduced in Fig. 4 for the case of a cubic crystal with three infrared active phonons [17]. The LO modes are represented by dashed horizontal lines. Only the frequency of the lowest TO branch goes to zero for $k \to 0$. The two other TO branches merge with the next lower LO mode for $k = 0$, as does the upper polariton branch. The relations (II.26) and (II.27) retain their validity beyond the special case considered here for non-cubic crystals, too, when the dispersion is only to be described for directions parallel to the principal crystal axes.

c) Uniaxial, Polyatomic Crystals

Here a new aspect arises. When the optical axis of the crystal points in the z-direction of an orthogonal coordinate system, the xy-plane is isotropic. Therefore the physical quantities are denoted according to their relation with respect to the optical axis:

$$\epsilon_1 = \epsilon_x = \epsilon_2 = \epsilon_y = \epsilon_\perp, \; \epsilon_3 = \epsilon_z = \epsilon_{\|}.$$

For the direction cosine we write

$$s_1^2 + s_2^2 = s_x^2 + s_y^2 = s_\perp^2 = \sin^2 \vartheta, \; s_3^2 = s_z^2 = s_{\|}^2 = \cos^2 \vartheta.$$

ϑ is the angle between the optical axis (z-direction) and the wave vector $\vec{k}$. From the general dispersion Eq. (II.22a) it then follows

$$(\epsilon_\perp - n^2) [\epsilon_\perp s_\perp^2 (\epsilon_{\|} - n^2) + \epsilon_{\|} s_{\|}^2 (\epsilon_\perp - n^2)] = 0. \qquad (II.28)$$

There are again two possible ways to fulfil this equation.

α) Ordinary Phonons and Polaritons. The condition $\epsilon_\perp - n^2 = 0$ leads to the dispersion relation

$$n^2 \equiv \frac{k^2 c^2}{\omega^2} = \epsilon_\perp (\omega). \qquad (II.29)$$

One can show that for this solution $\vec{E}$, $\vec{P}$, and $\vec{w}$ are perpendicular to $\vec{k}$ and the z-direction. Phonons and polaritons for which condition (II.29) is fulfilled are called "ordinary". Because of $\vec{w} \perp \vec{k}$ there are by definition only transverse ordinary phonons. In Eq. (II.29) the right-hand side is

$$\epsilon_\perp (\omega) = \epsilon_{\infty\perp} \prod_{i=1}^{m_\perp} \frac{\omega_{L\perp i}^2 - \omega^2}{\omega_{T\perp i}^2 - \omega^2} . \qquad (II.30)$$

$m_\perp$ is the number of infrared-active phonons when the infrared spectrum is observed with the polarization perpendicular to the optical axis. $\omega_{T\perp i}$ are the frequencies of the transverse phonons and $\omega_{L\perp i}$ those of the longitudinal phonons, both measured perpendicular to the optical axis. Because of the analogy between Eqs. (II.21) and (II.29), the behaviour of the disper-

101

sion curves of the ordinary phonons in uniaxial crystals is the same as that of the transverse phonons in cubic crystals. They appear as in Fig. 4, but with the addition of an index $\perp$ to every ω. It is remarkable that no direction cosine occurs in the dispersion Eq. (II.29) for the ordinary phonons. This means that the ordinary phonons show dispersion only with respect to the magnitude and not to the direction of the wave vector. This is not the case for the second possible fulfillment of Eq. (II.28).

β) *Extraordinary Phonons and Polaritons.* The dispersion equation for extraordinary phonons and polaritons is

$$\epsilon_\perp s_\perp^2 (\epsilon_\parallel - n^2) + \epsilon_\parallel s_\parallel^2 (\epsilon_\perp - n^2) = 0. \tag{II.31}$$

This equation gives with $s_\perp = \sin \vartheta$ and $s_\parallel = \cos \vartheta$

$$n^2 \equiv \frac{k^2 c^2}{\omega^2} = \frac{\epsilon_\parallel \epsilon_\perp}{\epsilon_\parallel \cos^2 \vartheta + \epsilon_\perp \sin^2 \vartheta}. \tag{II.32}$$

There is no simple orientation between the optical axis and $\vec{k}$ on the one hand and $\vec{E}$, $\vec{P}$, and $\vec{w}$ on the other hand for arbitrary directions of $\vec{k}$. It must be emphasized that the direction cosines appear in Eq. (II.32), and therefore the extraordinary phonons and polaritons show dispersion not only with respect to the magnitude, but also to the direction of the wave vector.

The situation is relatively easy to comprehend for the limiting cases $\vartheta = 0$ and $\vartheta = \pi/2$. $\vartheta = 0$ means that $\vec{k}$ and z have the same direction. Then $s_\perp = 0$ and $s_\parallel = 1$. From Eq. (II.31) it follows

$$\epsilon_\parallel (\epsilon_\perp - n^2) = 0. \tag{II.33}$$

Again there are two solutions. From $\epsilon_\parallel = 0$ we obtain the Lyddane-Sachs-Teller relation, now in the form

$$\prod_{i=1}^{m \parallel} \omega_{L\parallel i}^2 = \frac{\epsilon_{0\parallel}}{\epsilon_{\infty\parallel}} \prod_{i=1}^{m \parallel} \omega_{T\parallel i}^2. \tag{II.34}$$

Therefore LO phonons exist with the frequencies $\omega_{L\parallel i}$ independent of k.

The other solution $\epsilon_\perp - n^2 = 0$ leads to strictly transverse optical phonons. They degenerate with the ordinary phonons given by Eq. (II.29).

$\vartheta = \pi/2$ finally means that the wave vector $\vec{k}$ propagates perpendicularly to the optical axis. Then $s_\parallel = 0$ and $s_\perp = 1$. From Eq. (II.31) it then follows

$$\epsilon_\perp (\epsilon_\parallel - n^2) = 0. \tag{II.35}$$

$\epsilon_\perp = 0$ again leads to k-independent LO phonons with the frequencies determined by

$$\prod_{i=1}^{m\perp} \omega_{L\perp i}^2 = \frac{\epsilon_{0\perp}}{\epsilon_{\infty\perp}} \prod_{i=1}^{m\perp} \omega_{T\perp i}^2. \tag{II.36}$$

The further solution $\epsilon_\parallel - n^2 = 0$ describes the dispersion of another set of strictly transverse phonons which do not degenerate with the ordinary phonons because of the anisotropy of the crystal.

In all other cases with $0 < \vartheta < \pi/2$ no relations of the form $\epsilon = 0$ and $\epsilon - n^2 = 0$ can be derived from Eq. (II.32). This means that only for the limiting cases $\vartheta = 0$ and $\vartheta = \pi/2$ do pure LO and TO phonons exist. In all other directions the eigenvibrations have a more or less mixed character.

In order to obtain approximate relations for directions $0 < \vartheta < \pi/2$ Merten [18] and Loudon [19] considered a number of special cases where either the anisotropy of the crystal or the electric fields causing LO-TO splitting may be neglected.

B. Experimental Observation of Polaritons by Raman Scattering

The first experimental observation of polaritons was made in 1965 by Henry and Hopfield [20] in the cubic diatomic crystal gallium phosphide. The experimental arrangement must be chosen so as to yield small k vectors. This is the case for near forward scattering, namely when the angle φ between the wave vectors of the incident and scattered light, $\vec{k}_i$ and $\vec{k}_s$, respectively, becomes small. A device with which this may be achieved was described by Claus [21]. According to the dispersion curves (e.g. in Fig. 4) the Raman lines arising from polaritons are shifted towards smaller wave numbers for decreasing k values.

1. Zinc Oxide

ZnO was the first uniaxial crystal in which polariton dispersion was detected by Porto et al. [22] in 1966. ZnO is hexagonal and has the factor group C_{6v}. From the character table for the point group C_{6v} one can immediately deduce that the vibrations of species A_1 and E_1 are infrared-active, i.e. polar, and Raman-active. The vibrations of species E_2 on the other hand are nonpolar and exclusively Raman active. In an earlier paper Damen, Porto, and Tell [23] determined the phonon spectrum of zinc oxide by 90° Raman scattering. In this case the phonon wave vectors are much greater than those required for the observation of polariton dispersion. For small scattering angles φ the Raman line corresponding to the E_1 (TO) phonon at 407 cm^{-1} was observed to shift towards smaller wave numbers [22]. For ZnO the directional dispersion can be neglected. However, this is not the case for our next example.

2. α-Quartz

This uniaxial crystal belongs to the trigonal system (factor group D_3) and has 9 atoms in the elementary cell. Vibrations of species E are polar and Raman-active and polariton dispersion may be expected for the corresponding lines. The polariton spectrum was observed by Scott *et al.* [24] and the dispersion curves for the ordinary and the extraordinary phonons and polaritons were calculated by Merten [25].

3. Lithium Iodate

$LiIO_3$ is a uniaxial hexagonal crystal (factor group C_6). Vibrations of species A, E_1, and E_2 are allowed in the Raman effect, but only A and E_1 are infrared-active, therefore polariton dispersion is expected for the transverse phonons of these two species. The phonon and polariton spectra were investigated by Claus [26,27] and Otaguro *et al.* [28,29]. Here we want to show two series of spectra recorded by Claus.

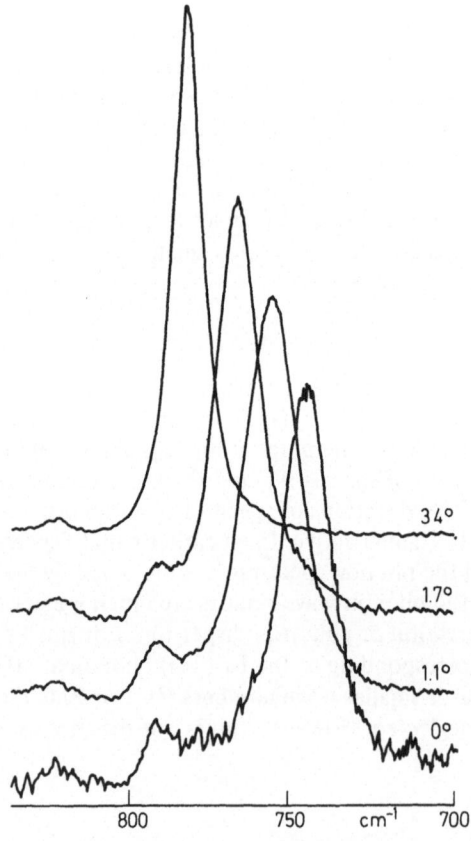

Fig. 5. Polariton associated with the A (TO) phonon at 795 cm^{-1} in LiIO$_3$. Wave vector triangles lie in the optically isotropic xy plane of the crystal [21]

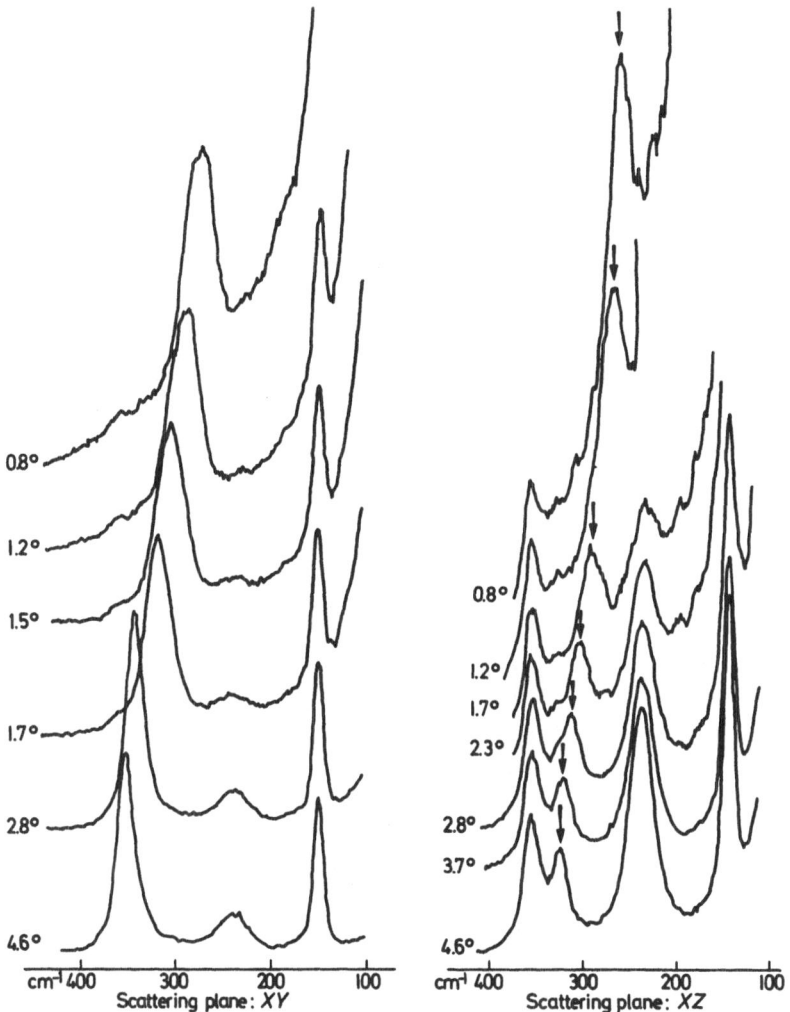

Fig. 6. Polariton associated with the A(TO) phonon at 354 cm^{-1} in LiIO$_3$. a) Scattering plane xy, b) scattering plane xz [35a)]

At 795 cm^{-1} LiIO$_3$ has a transverse A phonon. Fig. 5 shows the shift of the Raman line towards lower wave numbers for decreasing angles between the directions of the incident and scattered light. Because the wave vector triangle lies in the isotropic xy-plane, $\vec{k}$ is always perpendicular to the optical axis and no directional dispersion is to be expected, therefore the shift of the line is due only to the variation of the magnitude of the wave vector.

105

$LiIO_3$ is called a negative crystal, because the extraordinary refractive index is smaller than the ordinary one: $n_e - n_0 = 1.78 - 1.93 < 0$ (values for 514.5 nm). When the scattering geometry is chosen in such a way that the incident light is an extraordinary ray, a greater polariton shift is obtained because smaller k values can be realized, as in the case of ZnO.

Fig. 6 shows the scattering spectra of the polariton arising from the A phonon at 354 cm^{-1}. For the left series of spectra the scattering plane is the xy-plane, $i.e.$ no directional dispersion can occur. With decreasing angle the line is shifted towards smaller wave numbers. The right series was recorded with a scattering geometry in the xz-plane. The spectra look quite different because of the additional influence of directional dispersion, and at $\varphi = 4.6°$ the polariton line does not reach the position of the A phonon line at

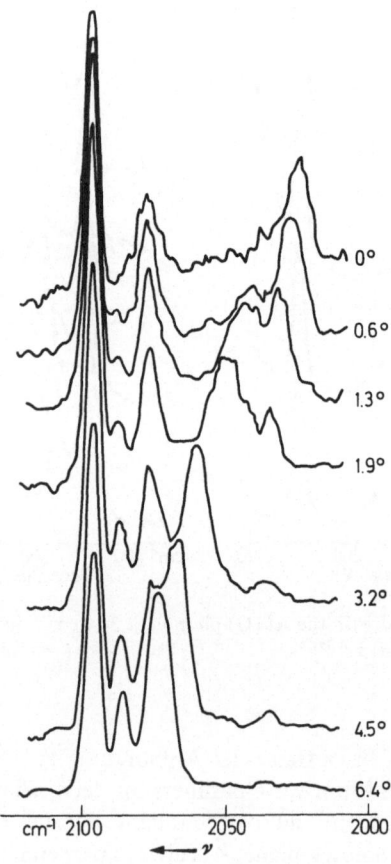

Fig. 7. "Crossing" of a polariton branch with a second-order phonon line in $K_3Cu(CN)_4$. The scattering angles are those inside the sample [31)]

354 cm^{-1}, as in the left series, but only the E$_1$ phonon at 328 cm^{-1} to which it is coupled. Otaguro et al. [29] have given a complete assignment of the phonon spectrum, so that a full comparison of theoretical dispersion curves with the experimental results is now possible.

4. Potassium Copper Cyanide

In this crystal an interesting effect occurs [30]. $K_3Cu(CN)_4$ is uniaxial (factor group C_{3v}), species A$_1$ and E are allowed in infrared absorption and in the Raman effect. Polariton behaviour was observed for the TO phonon at 2076 cm^{-1} of species A$_1$ which corresponds to an internal vibration of the cyanide group. Fig. 7 shows the spectra recorded with He-Ne laser excitation [31]. The scattering plane was xy, therefore no directional dispersion influences the results. With decreasing angle φ a line is shifted towards smaller wave numbers. When it approaches a weak second-order phonon line at 2035 cm^{-1}, the latter increases in intensity and begins to shift, while the original polariton line loses its intensity and finally comes to a standstill at the position of the second-order line. The exchange of intensities occurs in the same way as in Fermi resonance in molecular spectra. Analogous behaviour was observed in the same crystal for a polariton originating from an E phonon at 2082 cm^{-1}. Agranovich et al. [32] have quoted this phenomenon as experimental evidence for their theoretical considerations regarding Fermi resonance in molecular crystals.

5. Lithium Niobate

a) Upper Polariton Branch

The measurements reported above did not refer to the upper polariton branch which was first observed in LiNbO$_3$. We refer here to the branch having phonon character for small k values and approaching the photon dispersion line for k large. First, we wish to make some more general remarks. For scattering of light in a material medium, energy conservation leads to the relation $\omega_i = \omega_s + \omega$ for the Stokes case. For Raman scattering at large angles, ω is the frequency of a phonon. For polariton scattering, ω is the frequency of a coupled phonon-photon state. Finally the same relation is also valid for parametric luminescence. Here ω is the frequency of a photon; in this connection it is usually called the "idler" frequency. In the upper polariton branch a transition can be observed from Raman scattering at a phonon to polariton scattering and finally to parametric luminescence.

Apart from energy conservation, momentum conservation must also be fulfilled: $\vec{k}_i = \vec{k}_s + \vec{k}$. From this wave vector relation a short calculation for exact forward scattering, *i.e.* $\varphi = 0$, leads to the following condition for ω and k which must be fulfilled in order to make the scattering observable:

$$\omega = \frac{k}{n_s} - \frac{n_i - n_s}{n_s} \omega_i. \qquad (II.37)$$

n_i and n_s are the refractive indices for the incident and scattered light, respectively. Eq. (II.37) is a linear relation between ω and k as long as dispersion of n_s is neglected. $k = 0$ can be realized for a finite value of ω if $n_i < n_s$. LiNbO$_3$ is a negative crystal, *i.e.* $n_e - n_0 < 0$. Therefore the scattering geometry must be chosen in such a way that the incident ray is extraordinary and the scattered ordinary. In LiNbO$_3$ the numerical values for the refractive indices are so favourable that $k = 0$ for exciting light of a wavelength near 500 nm and for an ω value very close to the frequency of the highest LO phonon into which the upper polariton branch merges. Claus *et al.* [33] photoelectrically recorded this upper polariton branch for the first time and obtained good agreement between theory and experiment.

In connection with parametric luminescence, the upper polariton branch of LiNbO$_3$ was photographed earlier by Klyshko *et al.* [34]. The beam of an argon ion laser (488 nm) was incident perpendicular to the optical axis of their LiNbO$_3$ crystal. Behind the crystal a lens focused the forward scattered light on the entrance slit of a spectrograph. The light scattered in directions φ to the forward direction is then focused at different heights on the entrance slit, so that the slit-height dimension becomes a function of φ, and on the spectrogram an (ω, φ) diagram is obtained. The straight line representing the observation limit (II.37) for $\varphi = 0$ twice crosses the upper polariton branch. Between $\varphi = 0°$ and $4°$ two intersections are always obtained until the hyperbola [35] representing $\varphi = 4°$ merely touches the upper polariton branch, *i.e.* an elliptic figure is observed on the spectrogram [34].

b) Directional Dispersion

In LiNbO$_3$ Claus [36] has also investigated directional dispersion. Borstel and Merten [37] have calculated the directional dispersion of all optical phonons of LiNbO$_3$ on the basis of the assignment given by Kaminow and Johnston [38]. However, the dispersion curve coupling the E(TO) phonon at 582 cm^{-1} to an E(LO) phonon at 621 cm^{-1} could not be verified by experiment; instead this E(TO) was found to couple to the A$_1$ (TO) at 637 cm^{-1}. Therefore the assignment [38] of the E(LO) at 621 cm^{-1} and also that of some other phonons had to be revised. The experimental basis for this reassignment are series of spectra obtained by right-angle and backscattering experiments on LiNbO$_3$ samples cut at different angles to the optical axis [39]. The directional dispersion has proven to be a very sensitive test for the assignment of the phonon spectrum of a crystal.

C. Plasmons and Coupled States

1. Plasmons

These are quantized vibrations of conduction electrons in a metal or semiconductor [40]. The quantized energy levels of the collective longitudinal vibrations of the electron gas are quasi-particles in the sense of the definition given above and are called plasmons. The frequency of this longitudinal vibration, the plasma frequency ω_P, is given by [40]

$$\omega_P = \sqrt{\frac{4 \pi n e^2}{\epsilon_\infty m^*}}, \tag{II.38}$$

where n denotes the electron density, e the elementary charge, ϵ_∞ the optical dielectric constant, and m^* the effective mass of the electrons in the conduction band, which can be appreciably lower than the normal mass of the electron.

2. Coupled Plasmon-Phonon States

The coupling of plasmons with LO phonons leads to coupled plasmon-phonon states. For a cubic crystal with 2 atoms in the elementary cell and one infrared-active eigenfrequency ω_T, the frequency-dependent dielectric constant is according to Eq. (II.17)

$$\epsilon(\omega) = \epsilon_\infty \frac{\omega_L^2 - \omega^2}{\omega_T^2 - \omega^2}.$$

This relation holds for a completely isolating crystal. When conduction electrons are present, the so-called Drude-term has to be added on the right-hand side and we have

$$\epsilon(\omega) = \epsilon_\infty \frac{\omega_L^2 - \omega^2}{\omega_T^2 - \omega^2} - \epsilon_\infty \frac{\omega_P^2}{\omega\left(\omega + \frac{i}{\tau_0}\right)}. \tag{II.39}$$

ω_P is the plasma frequency defined by (II.38), τ_0 the mean collision relaxation time for the electrons which determines the damping, and i the imaginary unit. When damping is neglected (as for the phonons) we are left with

$$\epsilon(\omega) = \epsilon_\infty \left(\frac{\omega_L^2 - \omega^2}{\omega_T^2 - \omega^2} - \frac{\omega_P^2}{\omega^2} \right). \tag{II.40}$$

According to the general rule, LO phonons are obtained when $\epsilon = 0$. With (II.38) it follows from (II.40)

$$\frac{4\pi n e^2}{m^*} \cdot \frac{1}{\omega^2} = \epsilon_\infty \frac{\omega_L^2 - \omega^2}{\omega_T^2 - \omega^2}. \tag{II.41}$$

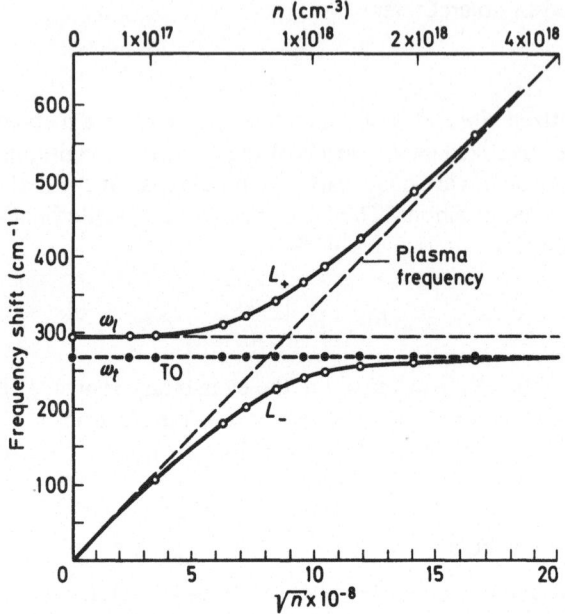

Fig. 8. Frequency shift of the Raman-scattered light in GaAs at room temperature as a function of the square root of the electron concentration [41]

We obtained earlier a very similar expression for the dispersion of the TO branches, namely

$$k^2 c^2 \; \frac{1}{\omega^2} = \epsilon_\infty \frac{\omega_L^2 - \omega^2}{\omega_T^2 - \omega^2} \; . \qquad (\text{II}.21)$$

Therefore the dispersion of the LO plasmon-phonon states is formally equivalent to the dispersion of the TO photon-phonon states, with $4\pi n e^2/m^*$ replacing $k^2 c^2$. When the plasmon-phonon frequency ω is plotted against $\sqrt{n}$ instead of k, dispersion curves for the LO modes are obtained which are similar to the polariton dispersion curves, the TO phonons showing no dispersion with $\sqrt{n}$.

For high electron densities n, scattering at the pure plasmon is obtained, whereas for relatively low values of n coupled plasmon-phonon states are observed. In Fig. 8 the dispersion curves for gallium arsenide measured by Mooradian et al. [41,42] are shown. The black dots represent the unshifted TO phonons and the circles the coupled modes designated L_+ and L_-. The calculated full curves agree very well with the measured values. The scattering at plasmons and coupled states was also measured [43] for InP, CdTe, and AlSb and calculated [44] for InSb.

110

3. Plasmaritons

Plasmon-phonon coupling represents mixing of two quasi-particles. The coupling of three quasi-particles has also been observed. The term plasmariton was used by Alfano [45] for a coupled state of a TO phonon and a "dressed photon", namely, a photon surrounded by an electron cloud (a coupled state of a plasmon and a photon). The quasi-particle "dressed photon" is also called a transverse plasmon. Because the coupled state of a photon and a TO phonon has been termed polariton, a plasmariton can also be regarded as coupled state of a plasmon and a polariton. Earlier [46] the term plasmariton was used in a more restricted sense, namely, when a partly transverse character of the plasmon is induced by an external magnetic field.

The corresponding dispersion curves are obtained by insertion into Eq. (II.21) for the transverse states of Eq. (II.39) for $\epsilon(\omega)$ including the Drude term. Disregarding damping, this gives

$$\frac{k^2 c^2}{\omega^2} = \epsilon_\infty \left(\frac{\omega_L^2 - \omega^2}{\omega_T^2 - \omega^2} - \frac{\omega_P^2}{\omega^2} \right) \tag{II.42}$$

for the case of a cubic, diatomic crystal. Patel and Slusher [46] discussed and evaluated this expression for GaAs. They found that the polariton dispersion curve is modified in two ways. Firstly the lower polariton branch does not go to $\omega = 0$ for $k = 0$. For an electron densitiy of $n = 2.9 \cdot 10^{17}$ cm^{-3} an ordinate value of $\omega = 160$ cm^{-1} is obtained. Secondly the wave number of the k-independent LO phonon is increased by about 10 cm^{-1}. The same authors [46] also investigated the influence of an external magnetic field of $8 \cdot 10^6$ A/m (100kOe). Shah et al. [46a] observed Raman scattering from polaritons and plasmaritons in CdS. McWhorter and Argyres [47] theoretically studied Raman scattering of magnetoplasma waves in semiconductors with special reference to GaAs.

D. Scattering at Landau Levels

Landau levels are the quantized orbitals of free particles in a crystal in an external magnetic field [40]. The motion of electrons in a magnetic field is governed by the Biot-Savart law. The energy levels of the circular orbitals of the electrons revolving with the cyclotron frequency are called Landau levels. The inelastic scattering of electromagnetic radiation in such a system, causing a transition from one Landau level into another, is called Landau-level Raman scattering. Theoretical studies of this effect were made by Wolff [48], Kelley and Wright [49], and Yafet [50]. Slusher et al. [51] confirmed the theoretical predictions through their measurements in InSb. Fig. 9 shows their results. The experimental arrangement is indicated in the upper part of the figure. The beam of a giant-pulse CO_2 laser with a peak power of 25 kW at a wavelength of 10.6 μm is focused by a BaF_2 lens into the InSb sample which is placed at

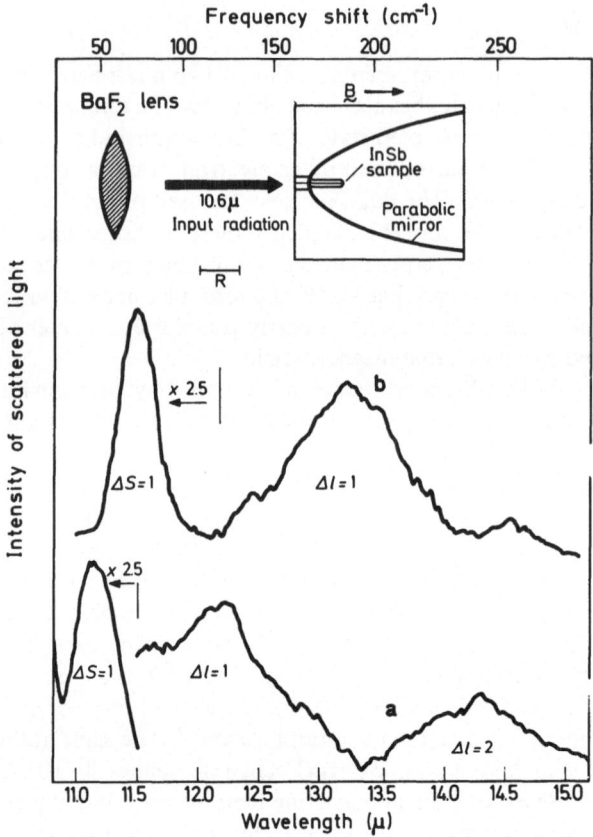

Fig. 9. Spectrum of light scattered at 90° from n-InSb ($n_e = 5 \cdot 10^{16}$ cm^{-3}). Notice the scale change in curve (a) for the intensity of the spin-flip scattered light

the focal point of a parabolic mirror. The scattered light is detected by a Ge-Cu detector cooled to 4.2 K. The lower spectrum (a) is recorded at a magnetic field of $2.10 \cdot 10^6$ A/m (26.2 kOe), the upper (b) at $2.94 \cdot 10^6$ A/m (26.7 kOe). The line designated $\Delta S = 1$ corresponds to the "spin-flip" of the electrons, the broad lines $\Delta l = 1$ and $\Delta l = 2$ are Landau-level transitions. l is the Landau-level quantum number. At higher magnetic fields the lines are shifted towards higher wave numbers. Further theoretical considerations were made by Wright et al. [52] and Makarov [53].

E. Light Scattering at Magnons

Internal and external magnetic fields also play a special role for the observation of light scattering at magnons, as the quanta of spin waves are called [40] - see Fig. 10.

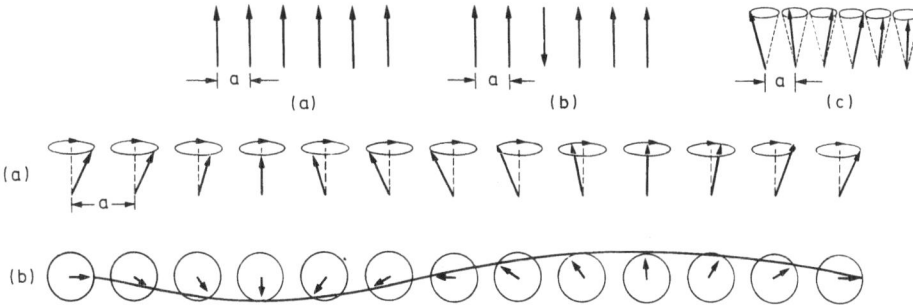

Fig. 10. Explanation of spin waves (see text) [40]

In a ferromagnetic material the spins are parallel within a Weiss domain. The left upper part of the figure shows the classical picture of the ground state of a ferromagnet. All spins are parallel. A possible energy level requiring a relatively high energy is drawn at top center: one spin is completely flipped. However, when the energy is not sufficient for a complete flipping, the spins can only deviate from their equilibrium (parallel) position by a small amount. Such a low-lying excitation is shown on the right. These elementary excitations are called spin waves. The spin vectors precess on cones and successive spins have a constant angle of phase shift. This is shown in the lower part of the figure showing one wavelength of a spin wave in a chain of spins (a) in perspective projection and (b) viewed from above.

1. Antiferromagnetic Magnons

There are also spin waves in antiferromagnetic materials in which the spins are paired with anti-parallel directions. Their quanta are called antiferromagnetic magnons. FeF_2 is antiferromagnetic. The infrared absorption corresponding to a two-magnon process was known to be situated at about 154 cm^{-1} at 5 K in this substance. Further, the position of the one-magnon line at 52.7 cm^{-1} near 0 K had been determined by antiferromagnetic resonance.

Porto et al. [54] investigated the Raman spectrum of a FeF_2 crystal with argon laser excitation at 488 nm. The phonon spectrum of the unaxial crystal FeF_2 could be fully explained on the basis of the factor group D_{4h}. The low-temperature Raman spectra were recorded by Fleury et al. [55]. The result in the low-frequency region is shown in Fig. 11. The Néel temperature of FeF_2 is 78.5 K, i.e. the substance is paramagnetic above this temperature and antiferromagnetic below it. With decreasing temperature of the sample a Raman line appears in the wing of the exciting line and shifts to higher wave num-

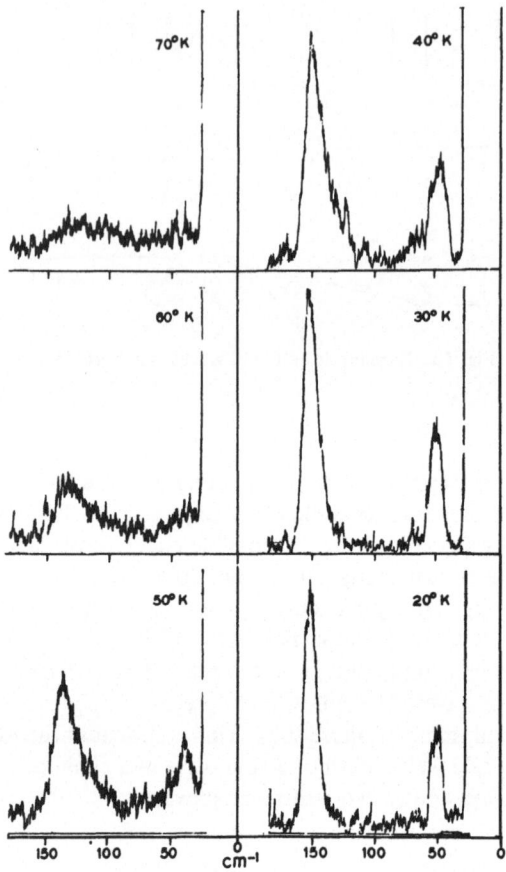

Fig. 11. Low frequency shift Raman Spectra in the (zy) scattering geometry for various temperatures in FeF_2 [55]

bers; at 20 K it has reached 50 cm^{-1}, in good agreement with the one-magnon line at 52.7 cm^{-1} known from antiferromagnetic resonance. This line was also observed in the anti-Stokes region. The temperature dependence of the wave number shift supports this assignment: when it is extrapolated towards the Néel temperature, it goes to $\omega = 0$ cm^{-1}, as expected from theoretical considerations. Moreover the polarization properties are in agreement with the selection rules for magnons derived by Shen and Bloembergen [56]. This means that the photons interact with the magnons indirectly through spin-orbit coupling and not directly through the magnetic field of the photon.

Fig. 11 also shows the gradual appearance of another line in the Raman spectrum at about 150 cm^{-1}. At 70 K it is weak and broad, at 20 K it becomes sharp and intense. The temperature dependence of its wave number is not so pronounced as for the one-magnon line. This indicates that the line corresponds to a second-oder process at the edge of BZ 1, where two magnons with $\omega \approx$ 77 cm^{-1} are simultaneously excited. The resulting wave vector $\vec{k} = \vec{k}_1 + \vec{k}_2$ is small when it is composed of two large wave vectors of nearly opposite direction (see Section A.4).

The intensities of these two lines pose interesting problems. According to theoretical calculations magnon scattering should be only one or two orders of magnitude less intense than vibrational Raman lines in liquids. Measurements show, however, that they are about four orders of magnitude weaker than the totally symmetric Raman line of benzene at 992 cm^{-1}. The increase in intensity with decreasing temperature is remarkable, as well as the fact that the two-magnon process has a higher scattering cross section than the one-magnon

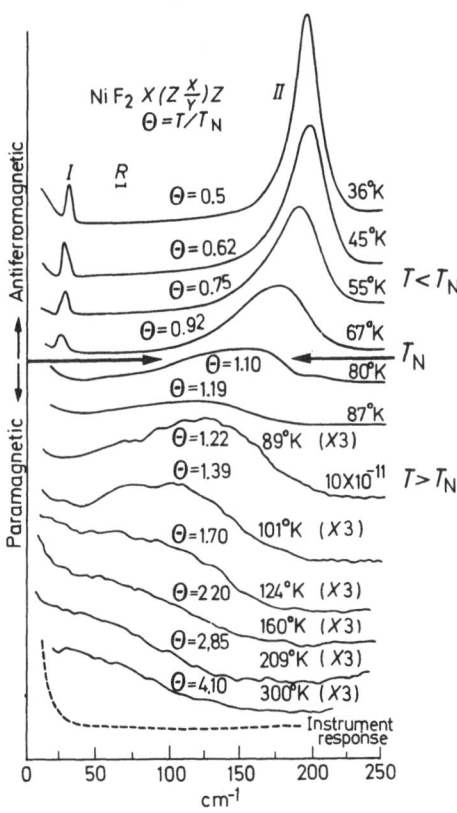

Fig. 12. Raman scattering from magnons in NiF$_2$ in the antiferromagnetic ($T < T_N$) and paramagnetic ($T > T_N$) phases [59]. Néel temperature T_N = 73 K. Threefold intensity scale above 88 K

process. Hutchings *et al.* [57] have investigated the dispersion curves of FeF_2 by inelastic neutron scattering and have compared their results with those of Raman spectroscopy. Antiferromagnetic magnons were also studied in MnF_2 [58]. One-magnon scattering is very difficult to detect in this case, because the Raman line is expected at 8 cm^{-1}, very close to the exciting line.

2. Paramagnons

Shen and Bloembergen [56] also predicted the existence of paramagnons and these were detected by Fleury *et al.* [59] in NiF_2. The spectra are shown in Fig. 12. NiF_2 is antiferromagnetic below 73 K. The one-magnon (I) and two-magnon (II) lines are clearly seen below this temperature. Above the Néel temperature the one-magnon line disappears, but two-magnon scattering is discernible up to 160 K. This means that spin waves also exist in the paramagnetic phase. These "paramagnons" have also been detected in rubidium nickel fluoride [60].

3. Ferromagnetic Magnons

These were found in $CdCr_4Se_4$ by Harbecke and Steigmeier [61] with He-Ne laser excitation.

4. Localized Magnons

These were observed by Oseroff and Pershan [62] in MnF_2 doped with Ni^{2+} and Fe^{2+}. They detected two-magnon scattering of the dopants and emphasize the excellent agreement between theory and experiment. It is obvious that the intensity problem was remarkable in this experiment. Magnon scattering is already very weak and here it has to be detected from a dopant.

III. Nonlinear Raman Effects

A. Stimulated Raman Scattering

Not long after the discovery of the stimulated Raman effect in liquids [63] it was also detected in single crystals [64], namely diamond, calcite, and α-sulfur. Only much later could it be shown that the effect can also be observed in crystal powders [65]. The stimulated Raman effect [99] is excited by giant-pulse lasers with a power of several MW. The strongest Raman lines of a substance are amplified until their intensity is of the same order of magnitude as that of the exciting line; furthermore second, third, etc. Stokes lines of the fundamentals in question are observed with twice, thrice, etc. the frequency shift.

Anti-Stokes stimulated Raman lines are emitted only in cones around the laser beam due to the wave-vector relation [66]

$$2\,\vec{k}_i = \vec{k}_s + \vec{k}_a,$$ (III.1)

where $\vec{k}_i$ is the wave vector of the incident photon and $\vec{k}_s$ and $\vec{k}_a$ those of the Stokes and anti-Stokes Raman photons, respectively. Quantitative agreement of the experimental data on calcite [67] and diamond [68] with relation (III.1) was obtained. This underlines the fact that the stimulated anti-Stokes radiation is generated through interaction of the laser beam and the Stokes radiation in the crystal. Various other aspects of stimulated Raman scattering in calcite were investigated by other authors [69-71].

1. Application to the Measurement of Phonon Lifetimes

Giordmaine and Kaiser [72] were the first to observe Stokes and anti-Stokes Raman scattering excited by a probe pulse from phonons in calcite which were generated by stimulated Raman scattering. The development of picosecond pulse lasers [73] allowed the direct measurement of optical phonon lifetimes. The much longer vibrational lifetime of hydrogen molecules in the gaseous state had been measured before [74]. Now Alfano and Shapiro [75] excited coherent optical phonons at 1086 cm^{-1} in calcite by stimulated Raman scattering of a 1.06 μm laser beam and subsequently measured the intensity of the Stokes Raman scattering of a probe pulse at 0.53 μm. A plot of the relative intensity at 0.562 μm, the Stokes Raman line, versus the variable delay time of the probe pulse gave an exponential decay with a lifetime of 8.5 ps at room temperature and 19 ps at 100 K, much greater values than those deduced from the Raman linewidths. A different result was obtained for the 1332 cm^{-1} phonon in diamond by Laubereau *et al.* [76] who used the phase-matched anti-Stokes Raman scattering of a delayed probe pulse as indicator of the phonon decay and measured lifetimes of 2.9 ps at room temperature and 3.4 ps at 77 K, in good agreement with the linewidth measured in the linear Raman effect.

2. Stimulated Polariton Scattering

This has aroused much interest because of the possibility of constructing tunable light sources in the infrared. Polariton behavior is only observed for polar, *i.e.* infrared active phonons. When polar phonons are coherently excited by stimulated Raman scattering, the stored vibrational energy can subsequently be emitted as infrared photons, so that stimulated polariton scattering will lead to tunable infrared emission. Before discussing the first experiments of this sort, we want to review briefly the results of linear Raman scattering at polaritons in LiNbO$_3$.

117

a) Polaritons in Lithium Niobate

Puthoff et al. [77)] found two polaritons associated with TO phonons of species A_1 with tuning ranges from 630 to 500 cm^{-1} and 250 to nearly 0 cm^{-1}. The Raman scattering at these extraordinary polaritons was used to study [36)] the directional dispersion of the phonons in LiNbO$_3$. Recently Winter and Claus [78)] also investigated the polaritons associated with the E phonons at 582 and 154 cm^{-1} by both photographic and photoelectric methods. These branches were important for the assignment of the phonon spectrum [39)].

The large scattering cross section of the extraordinary polaritons and their wide tuning range made tunable stimulated polariton scattering possible.

b) Tunable Stimulated Polariton Scattering in Lithium Niobate

Kurtz and Giordmaine [79)] were the first to observe stimulated Raman scattering at the polariton associated with the TO phonon at 630 cm^{-1} which was shifted to 497 cm^{-1} for 0° scattering excited with a Q-switched ruby laser. The corresponding phonon was also observed in this experiment. This can be explained by backward (180°) stimulated Raman scattering reflected from the laser resonator mirrors as confirmed by measurements of relative time of arrival at the spectrometer.

Stimulated Raman scattering at the lower extraordinary polariton branch was detected for scattering angles of 3.5° and 5° by Gelbwachs et al. [80)] in an

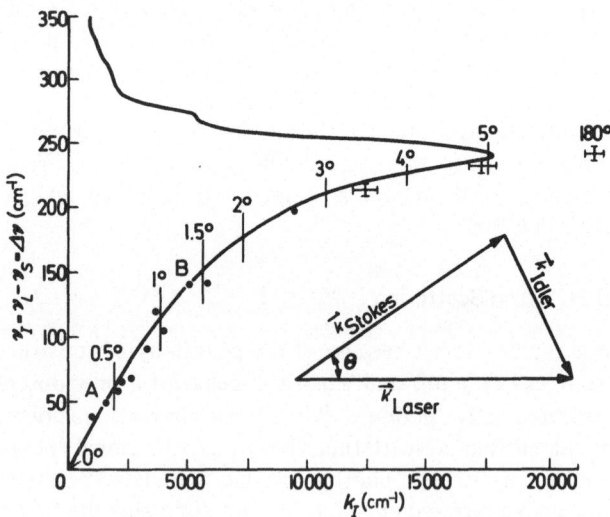

Fig. 13. Dispersion curve of LiNbO$_3$ for the polariton associated with the A_1 (TO) phonon at 248 cm^{-1}. Experimentally observed stimulated Raman scattering: crosses [80)] and dots [81)]

external optical resonator. The ruby laser beam was focused into the crystal at the corresponding angle to the resonator axis.

The limitation of the tuning range through the resonator was soon overcome by the same group [81] through the use of a longer crystal. The increased gain allowed them to use the polished plane parallel end faces of the crystal itself as the resonator and to accomplish the tuning through variation of the angle of incidence of the ruby laser beam.

Fig. 13 shows the polariton dispersion curve calculated from infrared reflectivity data. The inserted wave vector triangle visualizes the variation of the phonon wave vector ($\vec{k}_{idler}$) with scattering angle Θ. The k values realized for different angles are indicated. The measured data points [80,81] agree very well with the calculated curve. The sharp turn of this curve towards the ordinate axis is caused by the fact that pure spatial damping corresponding to IR-absorption has been considered and does not occur when pure temporal damping is taken into account [113,114]. On the right is indicated the frequency at which the TO phonon was observed for 180° scattering. Shifts smaller than 42 cm^{-1} could not be observed for very small scattering angles because the scattering cross section of this polariton decreases [82] for $k \to 0$ and, moreover, the gain is proportional to $\sqrt{\omega}$ for small ω and hence becomes too small.

The far-infrared emission of the "idler" frequency was also detected [81] with an InSb detector cooled with liquid He. The power of the pulses was estimated to be about 5 W. Their frequency was not directly measured but only inferred from energy conservation. Later measurements [83] gave a power from 0.25 W at 60 μm to 3 W at 200 μm and a linewidth of 0.1 to 0.5 cm^{-1} for the signal radiation.

c) Stimulated Polariton Scattering in Lithium Iodate

In LiIO$_3$ the polariton associated with the TO phonon at 795 cm^{-1} (see Fig. 5) has the largest scattering cross section when the incident light is polarized perpendicular to the optical axis (ordinary ray) [84]. In this scattering geometry at 0° stimulated Raman scattering from the polariton was observed at 753 cm^{-1}, *i.e.* the tuning range is only 40 cm^{-1} in this case. The second Stokes stimulated Raman line at 1506 cm^{-1} was also observed in this work [84]. Recently [85] a weak stimulated Raman line at 770 cm^{-1} was attributed to 2° scattering from this polariton in agreement with the dispersion curves determined by linear Raman scattering [27,29,35].

d) Difference Frequency Generation through Polariton Scattering

This was achieved in quartz [97] and gallium phosphide [98]. Two beams of different frequency generated through stimulated Raman scattering in appropriate liquids were directed to the crystal in directions calculated from the polariton dispersion curve and the wave vector relation. Polaritons were excited in the crystal through stimulated Raman amplification and the subsequent infrared emission was detected outside the crystal.

3. Stimulated Spin-Flip Raman Scattering

Another important source of tunable infrared radiation is stimulated spin-flip Raman scattering. Linear Raman scattering by the spin-flip of the conduction electrons had been studied earlier [51] together with Landau-level scattering (see Fig. 9). Patel and Shaw [86] succeeded in stimulating the spin-flip Raman process by excitation with a Q-switched CO_2 laser emitting 2 kW pulses at 10. 5915 μm. An InSb sample with an electron concentration of $n_e = 3 \cdot 10^{16}$ cm^{-3} was placed in a superconducting solenoid. When the magnetic field was varied from $3.8 \cdot 10^6$ to $8 \cdot 10^6$ A/m (48 to 100 kOe) the stimulated Raman emission shifted from 11.7 μm to 13 μm. Mooradian *et al.* [87] took advantage of the resonance-enhanced cross section for CO laser excitation for cw operation of the spin-flip Raman laser in magnetic fields between $1.4 \cdot 10^6$ and $4 \cdot 10^6$ A/m (17 and 50 kOe). The Stokes scattered frequency ω_s is given by

$$\omega_s = \omega_i - g \, \mu_B \, B, \qquad \qquad (III.2)$$

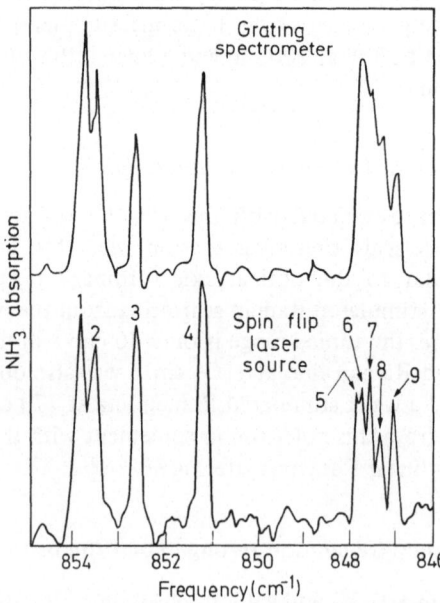

Fig. 14. Comparison of a part of the ν_2 band in the absorption spectrum of NH_3 taken with (a) the spin-flip laser [88] and (b) a conventional grating spectrometer [89]

where g is the g factor of the conduction electron, μ_B the Bohr magneton and B the magnetic field strength. The second Stokes stimulated line at $\Delta\omega = 2g\mu_B$ could also be detected [87]. Patel and Shaw [88] determined the linewidth of stimulated spin-flip Raman scattering to be $\lesssim 0.03$ cm^{-1} at 800 cm^{-1} and were able to record a part of the ν_2 absorption band of NH_3 by tuning the source without use of a spectrometer. Fig. 14 shows a comparison of their spectrum with one recorded by a conventional grating spectrometer [89].

Part of the water-vapour absorption band at 1800 cm^{-1} was recorded by Mellish *et al.* [90] in a similar way. They used a frequency-doubled CO_2 laser for excitation of the stimulated spin-flip scattering tunable in the range from 5.24 to 5.62 μm.

The resonance enhancement of the gain further studied by Brueck and Mooradian [91] was recently used by Patel [92] to extend the observation limit to magnetic fields to as low as $32 \cdot 10^3$ A/m (400 Oe), and he also observed the anti-Stokes stimulated Raman line.

In Table I we summarize the tuning ranges of stimulated emission obtained so far by stimulated polariton and spin-flip Raman scattering.

Table I. Tuning ranges of stimulated infrared emission

ω(cm^{-1})	λ(μm)	Reference
1. Polariton scattering		
42 – 250	240 – 40	80,81)
497 – 630	20 – 16	79)
753 – 795	13,3 – 12,6	84)
1506 – 1590	6,65 – 6,3	84)
2. Spin-flip scattering		
770 – 917	13 – 10.9	88)
1550 – 2000	6,45 – 5	87,90–92)

B. The Inverse Raman Effect

The inverse Raman effect was detected in liquids [93] soon after the discovery of the stimulated Raman effect. When a medium is irradiated simultaneously by intense monochromatic light from a giant-pulse laser and by a continuum, sharp absorption lines are observed on the anti-Stokes side of the laser line, and under special conditions also on the Stokes side [94]. McLaren and Stoicheff [95] used the intense fluorescence from a dye solution excited by frequency-doubled ruby laser radiation as the continuum together with the fundamental ruby laser pulse to investigate the inverse Raman spectra of some liquids and of a diamond crystal. The totally symmetric vibration was observed in the latter at 1332 cm^{-1}. The main potential of this effect lies in the investigation of short-lived species which is facilitated by the use of picosecond-pulses for excitation [96].

C. The Hyper-Raman Effect

This effect seems to be of major importance for the study of energy levels in crystals.

The intensity of Rayleigh scattering and the linear Raman effect is governed by the polarizability tensor $\alpha_{\rho\sigma}$ of a molecule and its derivatives with respect to the normal coordinates. When the electric field of the exciting radiation is very high, further terms in the expression for the induced dipole moment [104]

$$\mu_\rho = 4\pi\epsilon_0 (\sum_\sigma \alpha_{\rho\sigma}E_\sigma + \frac{1}{2}\sum_{\sigma\tau} \beta_{\rho\sigma\tau}E_\tau E_\sigma + \frac{1}{6}\sum_{\sigma\tau\nu} \gamma_{\rho\sigma\tau\nu}E_\nu E_\tau E_\sigma) + \ldots \qquad \text{(III.3)}$$

become significant, namely, the first and second hyperpolarizability $\beta_{\rho\sigma\tau}$ and $\gamma_{\rho\sigma\tau\nu}$, respectively. The hyperpolarizabilities lead to second and third harmonic light scattering at the frequencies $2\nu_i$ and $3\nu_i$ (hyper-Rayleigh scattering) and their derivatives with respect to the normal coordinates to nonlinear inelastic light scattering at $2\nu_i \pm \nu$ and $3\nu_i \pm \nu$ (hyper-Raman scattering), where ν_i denotes the frequency of the incident light and ν that of a molecular vibration, a phonon, or another elementary excitation.

The extremely weak light scattering due to the hyper-Raman effect was discovered in 1965 by Terhune, Maker, and Savage [100] in water, fused quartz, and carbon tetrachloride. The only investigation of a single crystal published to date was made on NH$_4$Cl (factor group O_h) [101]. In this case the potential of the hyper-Raman effect becomes apparent: because the selection rules for β-tensor scattering are different from those of infrared absorption and linear Raman scattering, an additional "silent" mode at 370 cm^{-1} could be observed. It corresponds to a librational motion of the NH$_4^+$ ions.

The selection rules for β hyper-Raman scattering were derived by Cyvin, Rauch, and Decius [102] and those for γ hyper-Raman scattering, which has not yet been detected experimentally, by Christie and Lockwood [103]. From their tables one can see that silent modes become β-active for such important point groups as C_6, D_6, C_{3v}, C_{6v}, C_{6h}, D_{3d}, O and O_h. Examples of additional γ activity can be found in the point groups C_{4v}, C_{6h}, D_{4h}, D_{6h}, and O_h. Long and Stanton [104] have derived a quantum-mechanical theory of the hyper-Raman effect which indicates several possibilities for resonance enhancement of hyper-Raman intensities. Iha and Woo [105] extended the theory of nonlinear light scattering in crystals to the case of two incident photons with frequencies ν_1 and ν_2.

IV. Conclusion

The use of lasers for the excitation of Raman spectra of solids has led to the detection of many new elementary excitations of crystals and to the observation of nonlinear effects. In this review we have tried to lead the reader to a basic understanding of these elementary excitations or quasi-particles, namely, phonons, polaritons, plasmons, plasmaritons, Landau levels, and magnons. Particular emphasis was placed upon linear and stimulated Raman scattering at polaritons, because the authors are most familiar with this part of the field and because it facilitates understanding of the other quasi-particles.

The investigation of elementary excitations in solids by Raman spectroscopy has developed very quickly in the last few years and will certainly lead to many more new results in the future. For example, the huge class of biaxial crystals has so far been avoided by many workers because of the difficulty of the experimental techniques required, but many interesting effects are to be expected from their study.

Intensity problems, second-order Raman spectra, Raman scattering from impurities, and phase transitions could not be reviewed in this article. These and other aspects of solid-state Raman spectroscopy were recently treated by Poulet and Mathieu [106], Pick [107], Scott [108], Barker and Loudon [109], Cowley [110] and Suschtschinskij [111], and further information may be found in the proceedings of the conferences on light scattering in solids [17,112].

V. Zusammenfassung

Der Einsatz von Lasern für die Erregung der Ramanspektren von Festkörpern führte zum Nachweis zahlreicher neuer Elementaranregungen in Kristallen und außerdem zur Beobachtung nichtlinearer Effekte bei hoher Leistungsdichte der Erregerstrahlung. Wir haben in diesem Artikel versucht, dem Leser ein grundsätzliches Verständnis dieser Elementaranregungen oder Quasi-Teilchen zu vermitteln, und zwar der Phononen, Polaritonen, Plasmonen, Plasmaritonen, Landau-Niveaus und Magnonen. Das Schwergewicht lag dabei auf der linearen und stimulierten Ramanstreuung an Polaritonen, da dieses Gebiet den Verfassern durch eigene Arbeiten vertraut ist und weil die dabei behandelten Erscheinungen das Verständnis der anderen Quasi-Teilchen erleichtern.

Die Untersuchung von Elementaranregungen in Festkörpern mit Hilfe der Ramanspektroskopie hat sich in den letzten Jahren sehr schnell entwickelt und wird in der Zukunft sicher noch viele neue Ergebnisse bringen. Beispielsweise wurde bisher die große Klasse der optisch zweiachsigen Kristalle infolge der größeren experimentellen Schwierigkeiten kaum behandelt, aber viele interessante Effekte sind von ihrer Untersuchung zu erwarten.

Auf Intensitätsprobleme, Ramanspektren zweiter Ordnung, Ramanstreuung an Fehlstellen und die Erscheinungen bei Phasenübergängen konnte in diesem Artikel nicht eingegangen werden. Diese und andere Aspekte der Ramanspektroskopie an Festkörpern wurden in letzter Zeit von Poulet und Mathieu [106], Pick [107], Scott [108], Barker und Loudon [109], Cowley [110] und Suschtschinskij [111] ausführlich behandelt. Weitere Informationsquellen sind die Tagungsberichte der Konferenzen über Lichtstreuung an Festkörpern [17,112].

Acknowledgement. We thank R. Claus for numerous suggestions and critical reading of the manuscript.

VI. References

[1] Smekal, A.: Naturwissenschaften *11*, 873 (1923).

[2] Haag, R.: Physik. Bl. *36*, 529 (1970).

[3] Brenig, W.: Naturwissenschaften *58*, 173 (1971).

[4] Rasetti, F.: Z. Physik *66*, 646 (1930).

[5] Hougen, J. T., Singh, S.: Phys. Rev. Letters *10*, 406 (1963).

[6] Koningstein, J. A.: J. Opt. Soc. Am. *56*, 1405 (1966); – Appl. Spectry. *22*, 438 (1968); – J. Chem. Phys. *51*, 1163 (1969). – Koningstein, J. A., Mortensen, O. S.: Phys. Rev. Letters *18*, 831 (1967); – J. Mol. Spectry. *27*, 343 (1968); – J. Chem. Phys. *48*, 3971 (1968); – Phys. Rev. *168*, 75 (1968). – Koningstein, J. A., Schaack, G.: J. Opt. Soc. Am. *60*, 755, 1110 (1970); – Phys. Rev. *B2*, 1242 (1970). – Grünberg, P., Koningstein, J. A.: J. Chem. Phys. *53*, 4584 (1970); – Chem. Phys. Letters *7*, 565 (1970); – Can. J. Chem. *49*, 2336 (1971). – Koningstein, J. A., in: Spectroscopy in inorganic chemistry (eds. C. N. R. Rao and J. R. Ferraro), Vol. 2. New York: Academic Press 1971. – Wadsack, R. L., Chang, R. K.: Solid State Commun. *10*, 45 (1972). – Koningstein, J. A., Mortensen, O. S., in: The Raman effect (ed. A. Anderson), Vol. 2. New York: Marcel Dekker 1972.

[7] Fast, H., Welsh, H. L., Lepard, D. W.: Can. J. Phys. *47*, 2879 (1969).

[8] Kittel, Ch.: Introduction to solid state physics. New York-London-Sydney: J. Wiley & Sons 1967.

[9] Hellwege, K. H.: Einführung in die Festkörperphysik I. Berlin-Heidelberg-New York: Springer 1968.

[10] Süssmann, G.: Z. Naturforsch. *11a*, 1 (1956); – Z. Naturforsch. *13a*, 1 (1958).

[11] Huang, K.: E. R. A. Report L/T *1950*, 239; – Proc. Roy. Soc. A *208*, 352 (1951).

[12] Born, M., Huang, K.: Dynamical theory of crystal lattices. Oxford: Clarendon Press 1954.

[13] Hopfield, J. J.: Phys. Rev. *112*, 1555 (1958).

[14] Merten, L.: Z. Naturforsch. *22a*, 359 (1967).

[15] Kurosawa, T.: J. Phys. Soc. Japan *16*, 1298 (1961).

[16] Barker, A. S.: Phys. Rev. *136*, 1290 (1964).

[17] Wright, G. B.: Light scattering spectra of solids. Berlin-Heidelberg-New York: Springer 1969.

18) Merten, L.: Z. Naturforsch. *16a*, 447 (1961).
19) Loudon, R.: Advan. Phys. *13*, 423 (1964).
20) Henry, C. H., Hopfield, J. J.: Phys. Rev. Letters *15*, 964 (1965).
21) Claus, R.: Rev. Sci. Instr. *42*, 341 (1971).
22) Porto, S. P. S., Tell, B., Damen, T. C.: Phys. Rev. Letters *16*, 450 (1966).
23) Damen, T. C., Porto, S. P. S., Tell, B.: Phys. Rev. *142*, 570 (1966).
24) Scott, J. F., Porto, S. P. S.: Phys. Rev. *161*, 903 (1967). – Scott, J. F., Cheesman, L. E., Porto, S. P. S.: Phys. Rev. *162*, 834 (1967).
25) Merten, L.: Z. Naturforsch. *24a*, 1878 (1969).
26) Claus, R., Schrötter, H. W., Hacker, H. H., Haussühl, S.: Z. Naturforsch. *24a*, 1733 (1969).
27) Claus, R.: Z. Naturforsch. *25a*, 306 (1970).
28) Otaguro, W., Arguello, C. A., Porto, S. P. S.: Phys. Rev. *B1*, 2818 (1970).
29) Otaguro, W., Wiener-Avnear, E., Arguello, C. A., Porto, S. P. S.: Phys. Rev. *B4*, 4542 (1971).
30) Claus, R.: Phys. Letters *31a*, 299 (1970).
31) Claus, R., Schrötter, H. W.: Opt. Commun. *2*, 105 (1970).
32) Agranovitsch, B. M., Lalov, I. I.: Soviet Phys.-Solid State *13*, 859 (1971).
33) Claus, R., Borstel, G., Merten, L.: Opt. Commun. *3*, 17 (1971).
34) Klyshko, D. N., Penin, A. N., Polkovnikov, B. F.: JETP Letters *11*, 5 (1970). – Krindach, D. P., Krol, L. M.: Opt. Spectr. *30*, 73 (1971).
35) Claus, R.: Phys. Status Solidi (b) *50*, 11 (1972).
35a) Claus, R.: Fachberichte zur 35. Physikertagung, p. 95. Stuttgart: Teubner 1970.
36) Claus, R., Schrötter, H. W., in: Proceedings of the International Conference on Light Scattering Spectra of Solids. Paris: Flammarion Sciences 1971.
37) Borstel, G., Merten, L.: Z. Naturforsch. *26a*, 653 (1971).
38) Kaminov, J. P., Johnston, W. D., jr.: Phys. Rev. *160*, 519 (1967); – Phys. Rev. *168*, 1045 (1968).
39) Claus, R., Borstel, G., Wiesendanger, E., Steffan, L.: Z. Naturforsch. *27a*, 1187 (1972).
40) Kittel, Ch.: Quantum theory of solids. New York: J. Wiley & Sons 1967.
41) Mooradian, A., Mc Whorter, A. L.: Phys. Rev. Letters *19*, 849 (1967).
42) Mooradian, A., Wright, G. B.: Phys. Rev. Letters *16*, 999 (1966).
43) Mooradian, A., et al., in: Light scattering spectra of solids (ed. G. B. Wright), pp. 285, 297. Berlin-Heidelberg-New York: Springer 1969.
44) Wolff, P. A., in: Light scattering spectra of solids (ed. G. B. Wright), p. 273. Berlin-Heidelberg-New York: Springer 1969.
45) Alfano, R. R.: J. Opt. Soc. Am. *60*, 66 (1970).
46) Patel, C. K. N., Slusher, R. E.: Phys. Rev. Letters *22*, 282 (1969).
46a) Shah, J., Damen, T. C., Scott, J. F., Leite, R. C. C.: Phys. Rev. *B3*, 4238 (1971).
47) McWhorther, A. L., Argyres, P. N., in: Light scattering spectra of solids (ed. G. B. Wright), p. 325. Berlin-Heidelberg-New York: Springer 1969.
48) Wolff, P. A.: Phys. Rev. Letters *16*, 225 (1966).
49) Kelley, P. L., Wright, G. B.: Bull. Am. Phys. Soc. *11*, 812 (1966).
50) Yafet, Y.: Phys. Rev. *152*, 858 (1966).
51) Slusher, R. E., Patel, C. K. N., Fleury, P. A.: Phys. Rev. Letters *18*, 77 (1967). – Patel, C. K. N., Slusher, R. E.: Phys. Rev. *167*, 413 (1968); – Phys. Rev. *177*, 1200 (1969); – Bull. Am. Phys. Soc. *13*, 480 (1968).
52) Wright, G. B., Kelley, P. L., Groves, S. H., in: Light scattering spectra of solids (ed. G. B. Wright), p. 335. Berlin-Heidelberg-New York: Springer 1969.
53) Makarov, V. P., in: Light scattering spectra of solids (ed. G. B. Wright), p. 345. Berlin-Heidelberg-New York: Springer 1969; – Soviet Phys. JETP *55*, 704 (1968).

54) Porto, S. P. S., Fleury, P. A., Damen, T. C.: Phys. Rev. *154*, 522 (1967).
55) Fleury, P. A., Porto, S. P. S., Cheesman, L. E., Guggenheim, H. J.: Phys. Rev. Letters *17*, 84 (1966).
56) Shen, Y. R., Bloembergen, N.: Phys. Rev. *143*, 372 (1966).
57) Hutchings, M. T., Rainford, B. D., Guggenheim, H. J.: Solid State Phys. *3*, 307 (1970).
58) Fleury, P. A., Porto, S. P. S., Loudon, R.: Phys. Rev. Letters *18*, 658 (1967). − Fleury, P. A., Loudon, R.: Phys. Rev. *166*, 514 (1968).
59) Fleury, P. A., in: Light scattering spectra of solids (ed. G. B. Wright), p. 185. Berlin-Heidelberg-New York: Springer 1969; − Bull. Am. Phys. Soc. *12*, 420 (1967).
60) Fleury, P. A., Worlock, J. M., Guggenheim, H. J.: Phys. Rev. *185*, 738 (1969).
61) Harbeke, G., Steigmeier, E., in: Light scattering spectra of solids (ed. G. B. Wright), p. 221. Berlin-Heidelberg-New York: Springer 1969.
62) Oseroff, A., Pershan, P. S., in: Light scattering spectra of solids (ed. G. B. Wright), p. 223. Berlin-Heidelberg-New York: Springer 1969.
63) Woodbury, E. J., Ng, W. K.: Proc. I. R. E. *50*, 2367 (1962). − Eckhardt, G., Hellwarth, R. W., McClung, F. J., Schwarz, S. E., Weiner, D., Woodbury, E. J.: Phys. Rev. Letters *9*, 455 (1962).
64) Eckhardt, G., Bortfeld, D. P., Geller, M.: Appl. Phys. Letters *3*, 137 (1963).
65) Zubov, V. A., Peregudov, G. V., Sushchinskij, M. M., Chirkov, V. A., Shuvalov, I. K.: JETP Letters *5*, 150 (1967). − Schrötter, H. W.: Naturwissenschaften *54*, 513 (1967).
66) Terhune, R. W.: Bull. Am. Phys. Soc. *8*, 359 (1963). − Garmire, E., Pandarese, F., Townes, C. H.: Phys. Rev. Letters *11*, 160 (1963).
67) Chiao, R., Stoicheff, B. P.: Phys. Rev. Letters *12*, 290 (1964).
68) McQuillan, A. K., Clements, W. R. L., Stoicheff, B. P.: Phys. Rev. *A1*, 628 (1970).
69) Gandrud, W. B., Moos, H. W.: J. Appl. Phys. *38*, 421 (1967).
70) Bisson, G., Mayer, G.: J. Phys. *29*, 97 (1968).
71) Ataev, B. M., Lugovoi, V. N.: JETP Letters *7*, 38 (1968); − Opt. Spectr. *26*, 567 (1969); − Opt. Spectr. *27*, 380 (1969); − Soviet Phys.-Solid State *10*, 1566 (1969).
72) Giordmaine, J. A., Kaiser, W.: Phys. Rev. *144*, 676 (1966).
73) DeMaria, A. J., Stetser, D. A., Heynau, H.: Appl. Phys. Letters *8*, 174 (1966).
74) DeMartini, F., Ducuing, J.: Phys. Rev. Letters *17*, 117 (1966).
75) Alfano, R. R., Shapiro, S. L.: Phys. Rev. Letters *26*, 1247 (1971).
76) Laubereau, A., von der Linde, D., Kaiser, W.: Phys. Rev. Letters *27*, 802 (1971).
77) Puthoff, H. E., Pantell, R. H., Huth, B. G., Chacon, M. A.: J. Appl. Phys. *39*, 2144 (1968).
78) Winter, F. X., Claus, R.: Opt. Commun., *6*, 22 (1972).
79) Kurtz, S. K., Giordmaine, J. A.: Phys. Rev. Letters *22*, 192 (1969).
80) Gelbwachs, J., Pantell, R. H., Puthoff, H. E., Yarborough, J. M.: Appl. Phys. Letters *14*, 258 (1969).
81) Yarborough, J. M., Sussman, S. S., Puthoff, H. E., Pantell, R. H., Johnson, B. C.: Appl. Phys. Letters *15*, 102 (1969).
82) Obukhovskii, V. V., Ponath, H., Strizhevskii, V. L.: Phys. Status Solidi *41*, 847 (1970).
83) Johnson, B. C., Puthoff, H. E., Soo Hoo, J., Sussman, S. S.: Appl. Phys. Letters *18*, 181 (1971).
84) Schrötter, H. W.: Z. Naturforsch. *26a*, 165 (1971).
85) Amzallag, E., Chang, T. S., Johnson, B. C., Pantell, R. H., Puthoff, H. E.: J. Appl. Phys. *42*, 3251 (1971).
86) Patel, C. K. N., Shaw, E. D.: Phys. Rev. Letters *24*, 451 (1970).
87) Mooradian, A., Brueck, S. R. J., Blum, F. A.: Appl. Phys. Letters *17*, 481 (1970).
88) Patel, C. K. N., Shaw, E. D.: Phys. Rev. *B3*, 1279 (1971).
89) Mould, H. M., Price, W. C., Wilkinson, G. R.: Spectrochim. Acta *13*, 313 (1959).

90) Mellish, R. G., Dennis, R. B., Allwood, R. L.: Opt. Commun. *4*, 249 (1971).
91) Brueck, S. R. J., Mooradian, A.: Phys. Rev. Letters *28*, 161 (1972).
92) Patel, C. K. N.: Appl. Phys. Letters *19*, 400 (1971).
93) Jones, W. J., Stoicheff, B. P.: Phys. Rev. Letters *13*, 657 (1964).
94) Dumartin, S., Oksengorn, B., Vodar, B.: Compt. Rend. *261*, 3767 (1965).
95) McLaren, R. A., Stoicheff, B. P.: Appl. Phys. Letters *16*, 140 (1970).
96) Alfano, R. R., Shapiro, S. L.: Chem. Phys. Lett. *8*, 631 (1971).
97) Biraud-Laval, S., Chartier, G.: Phys. Lett. *30A*, 177 (1969).
98) de Martini, F.: Phys. Lett. *30A*, 319 (1969); Phys. Rev. *B4*, 4556 (1971).
99) Bloembergen, N.: Am. J. Phys. *35*, 989 (1967). − Schrötter, H. W.: Naturwissenschaften *54*, 607 (1967). − Lallemand, P., in: The Raman effect (ed. A. Anderson), Vol. 1. New York: Marcel Dekker 1971.
100) Terhune, R. W., Maker, P. D., Savage, C. M.: Phys. Rev. Letters *14*, 681 (1965).
101) Savage, C. M., Maker, P. D.: Appl. Opt. *10*, 965 (1971).
102) Cyvin, S. J., Rauch, J. E., Decius, J. C.: J. Chem. Phys. *43*, 4083 (1965).
103) Christie, J. H., Lockwood, D. J.: J. Chem. Phys. *54*, 1141 (1971).
104) Long, D. A., Stanton, L.: Proc. Roy. Soc. A *318*, 441 (1970).
105) Iha, S. S., Woo, J. W. F.: Nuovo Cimento *2B*, 167 (1971).
106) Poulet, H., Mathieu, J. P.: Spectres de vibration et symétrie des cristeaux. Paris: Gordon & Breach 1970.
107) Pick, R. M.: Advan. Phys. *19*, 269 (1970).
108) Scott, J. F.: Am. J. Phys. *39*, 1360 (1971).
109) Barker, Jr., A. S., Loudon, R.: Rev. Modn. Phys. *44*, 18 (1972).
110) Cowley, R. A., in: The Raman effect (ed. A. Anderson), Vol. 1. New York: Marcel Dekker 1971.
111) Sushchinskij, M. M.: Spektry kombinatsionnogo rasseyaniya molekul i kristallov. Moskva: Izdatelstvo Nauka 1969; − Raman spectra of molecules and crystals. New York, Jerusalem, London: Israel Program for Scientific Translations 1972.
112) Balkanski, M. (ed.): Light scattering in solids. Paris: Flammarion Sciences 1971.
113) Alfano, R. R., Giallorenzi, T. G.: Opt. Commun. *4*, 271 (1971).
114) Merten, L., Borstel, G.: Z. Naturforsch. *27a*, in press.

Received May 8, 1972

Topics in Current Chemistry

Fortschritte der chemischen Forschung

Vol. 24: Electronic Structure of Organic Compounds

12 fig. 57 pp. 1971
DM 18,—; US $5.80
ISBN 3-540-05540-1

Contents
H. Fischer: Chemically Induced Dynamic Nuclear Polarization.
J.-F. Labarre, F. Crasnier: Critique of the Notion of Aromaticity.

Vol. 25: Catalysis

26 fig. 157 pp. 1972
DM 48,—; US $15.30
ISBN 3-540-05542-8

Contents
J. Manassen: The Catalytic Activity of Organic and Metallo-Organic Compounds in Heterogeneous Systems. R. L. Banks: Catalytic Olefin Disproportionation.
W. Strohmeier: Homogene katalytische Hydrierung mit Komplexverbindungen der 8. Gruppe.
G.-M. Schwab: Elektronik der Trägerkatalysatoren.
F. Steinbach: Heterogeneous Photocatalysis.

Vol. 26: Inorganic and Analytical Chemistry

6 fig. 115 pp. 1972
DM 36,—; US $11.50
ISBN 3-540-05589-4

Contents
J. L. Margrave, K. G. Sharp, P. W. Wilson: The Dihalides of Group IVB Elements.
A. Meller: The Chemistry of Iminoboranes.
G. D. Christian: Atomic Absorption Spectroscopy for the Determination of Elements in Medical Biological Samples.

Vol. 27: Nonaqueous Chemistry

46 fig. 190 pp. 1972
DM 48,—; US $15.30
ISBN 3-540-05663-7

Contents
B. Kratochvil, H. L. Yeager: Conductance of Electrolytes in Organic Solvents.
V. Gutmann: Ionic and Redox Equilibria in Donor Solvents.
S. L. Smith: Solvent Effects and NMR Coupling Constants.

■ **Prospectus on request**

Vol. 28: π Complexes of Transition Metals

11 fig. 184 pp. 1972
DM 38,—; US $12.10
ISBN 3-540-05728-5

Contents
G. Häfelinger: Theoretical Considerations for Cyclic (pd) π Systems.
J. Tsuji: Organic Synthesis by Means of Transition Metal Complexes: Some General Patterns.
L. D. Pettit, D. S. Barnes: The Stability and Structures of Olefin and Acetylene Complexes of Transition Metals.
H. Werner: Ringliganden-Verdrängungsreaktionen von Aromaten-Metall-Komplexen.

Vol. 29: Automation in Analytical Chemistry

32 fig. 106 pp. 1972
DM 28,—; US $8.90
ISBN 3-540-05758-7

Contents
P. P. Fietzek, K. Kühn: Automation of the Sequence Analysis by Edman Degradation of Proteins and Peptides.
H. Clever: Der Analysenautomat DSA-560.
H. Krech: Ein Analysenautomat aus Bausteinen, die Braun-Systematik.

W. Marks: Der Technicon Autoanalyzer.
F. Oehme: Titrierautomaten zur Betriebskontrolle.

Vol. 30: Nuclear Quadrupole Resonance

23 fig. 176 pp. 1972
DM 42,—; US $13.40
ISBN 3-540-05781-1

Contents
A. Weiss: Crystal Field Effects in Nuclear Quadrupole Resonance.
L. Guibé: Nitrogen Quadrupole Resonance Spectroscopy.
W. Zeil: Bestimmung der Kernquadrupolkopplungskonstanten aus Mikrowellenspektren.
E. A. C. Lucken: Nuclear Quadrupole Resonance; Theoretical Interpretation.

Vol. 31: Stereo- and Theoretical Chemistry

48 fig. 142 pp. 1972
DM 38,—; US $12.10
ISBN 3-540-05841-9

Contents
D. J. Cram, J. M. Cram: Stereochemical Reaction Cycles. A. Pullman: Quantum Biochemistry at the All- or Quasi-All-Electrons Level.
K. F. Freed: The Theory of Radiationless Processes in Polyatomic Molecules.

Springer-Verlag Berlin · Heidelberg · New York

London · München · Paris · Sydney · Tokyo · Wien

Springer-Verlag
Berlin · Heidelberg · New York
London München Paris Sydney Tokyo Wien

H. Determann
Gel Chromatography

Gel Filtration · Gel Permeation
Molecular Sieves A Laboratory Handbook

By Dr. phil. nat.
Helmut Determann,
Privatdozent,
Institut für
Organische Chemie
der Universität
Frankfurt am Main

From the Book Reviews of the 1st Edition:
Dr. Determann's book is an up-to-date, complete and reliable guide to this technique, to which it supplies an introduction. No laboratory concerned with the separation and characterization of organic compounds of moderate or high molecular weight can afford to be without it.

Chemistry in Britain

Translation
(by Dr. Erhard Gross,
Bethesda, Md., and
Dr. John M. Harkin,
Madison, Wis.,)
of the German
Edition of Determann
„Gelchromatographie"

Second edition
With 40 figures
XII, 202 pages
1969
Cloth DM 32, –
US $ 10.20

From the Book Reviews of the German Edition:
The method of gel chromatography has within a short space of time become a valuable technique used in many chemical and medical laboratories. Its wide scope explains why most reviews describe only parts of the field, or fail to give a comprehensive account of experimental details. The present book closes this gap. Five chapters describe the present state of gel chromatography. The practical section which describes the gels, the apparatus, and the experimental techniques will be most useful to the novice. – In the theoretical section, the attempt by the author to compare critically the different postulates about gel chromatography will stimulate the reader to reflect our ideas concerning the method and to carry out further experiments. The detailed descriptions and the numerous literature references make this book a valuable guide for all faced by fresh problems in this field. It is to be hoped that this book will receive the attention it deserves by workers engaged in the study of those disciplines in which this technique can be used.

Angewandte Chemie, International Edition

In kritischen Übersichten werden in dieser Reihe Stand und Entwicklung aktueller chemischer Forschungsgebiete beschrieben. Sie wendet sich an alle Chemiker in Forschung und Industrie, die am Fortschritt ihrer Wissenschaft teilhaben wollen.

In der Regel werden nur Beiträge veröffentlicht, die ausdrücklich angefordert worden sind. Schriftleitung und Herausgeber sind aber für ergänzende Anregungen und Hinweise jederzeit dankbar. Manuskripte können in den „Fortschritten der chemischen Forschung" in Deutsch oder Englisch veröffentlicht werden.

Jeder Band der Reihe ist einzeln käuflich.

This series presents critical reviews of the present position and future trends in modern chemical research. It is addressed to all research and industrial chemists who wish to keep abreast of advances in their subject.

As a rule, contributions are specially commissioned. The editors and publishers will, however, always be pleased to receive suggestions and supplementary information. Papers are accepted for "Topics in Current Chemistry" in either German or English.

Any volume of the series may be purchased separately.